GW01191331

THE CHINA CLIPPERS

Basil Lubbock was born on 9 September 1876, the son of Alfred Lubbock. He was educated at Eton. In 1912 he married Dorothy Mary. He went to Canada in 1897 and went over the Chilcoot Trail into Klondyke in the second year of the gold excitement. On leaving Klondyke, he joined the crew of a four-mast barque as an ordinary seaman and came home round the Horn. During the Boer War he held a commission in Menne's Scouts, a South African corps. Between October 1914 and April 1919, he held a Territorial Commission in the Third Wessex Brigade of the RAF. He was a member of the Nautical Research Society and wrote many highly acclaimed nautical books. He died on 3 September 1944.

The China Clippers is the history of the celebrated clipper ships of the nineteenth century. Built for speed rather than carrying capacity, the early clippers were ideal for use in the opium trade between 1830–75, and their captains stood to make a fortune from every successful run. As well as cutting a dash as slavers or pirate ships in the Indies, the great China tea trade was another area where the clippers came into their own. The author has unearthed many an old log book and here tells of some of the incidents relating to the exciting tea races between 1859–72 before the advent of steam. The frequent hazards of typhoons, treacherous currents, pirate ships and ship wrecks commanded fine performances of seamanship and daring from the best captains. This fascinating book combines an exciting historical record with a universal joy of the sea.

Also available in the Century Seafarers series:

1. THE BIRD OF DAWNING by John Masefield
2. SAILING ALONE AROUND THE WORLD by Joshua Slocum
3. NIGHT WATCHES by W. W. Jacobs
4. A VOYAGE IN THE SUNBEAM by Lady Brassey
5. 'RACUNDRA'S' FIRST CRUISE by Arthur Ransome

THE CHINA CLIPPERS

BASIL LUBBOCK

Introduction by Eric Newby

CENTURY PUBLISHING
LONDON

HIPPOCRENE BOOKS INC.
NEW YORK

First published in Great Britain in 1914 by
Brown, Son & Ferguson Ltd

This edition published in 1984 by
Century Publishing Co. Ltd,
Portland House, 12–13 Greek Street, London W1V 5LE

Published in the United States of America by
Hippocrene Books Inc.
171 Madison Avenue
New York NY 10016

ISBN 0 7126 0341 7

Cover painting by courtesy of
Sotheby Parke Bernet & Co., London

Reprinted in Great Britain by
Richard Clay (The Chaucer Press) Ltd
Bungay, Suffolk

Dedication

Dedicated to

" . . . The sailor of the sail, breed of the oaken heart,
Who drew the world together and spread our race apart,
Whose conquests are the measure of thrice the ocean's girth,
Whose trophies are the nations that necklace half the earth.
Lord of the bunt and gasket, and master of the yard
To whom no land was distant to whom no sea was barred."

INTRODUCTION

The first book I ever read by Basil Lubbock—and subsequently I read all of them and for a time owned most of them—was *Round the Horn Before the Mast*. I was then sixteen. It was heady stuff. It describes how the author, a tall, tough Old Etonian who had taken part in the Gold Rush of 1898 to the Yukon, signed on in 1899 aboard a four-masted iron barque in San Francisco, having failed to make his fortune in the Klondyke, gave his address as "Bachelors' Club, Piccadilly" and worked his passage home to Liverpool as an ordinary seaman before the mast, arriving in time to set off for the Boer War. It is a splendid book. By far the best account of life in the fo'c'sle of a big British sailing ship at that period, and to me it will always be particularly memorable because nearly forty years after the voyage which its author describes, it inspired me to persuade my father to apprentice me in a Cape Horn windjammer engaged in the Australian grain trade. He didn't take much persuading but it did cost him fifty pounds to have me indentured, and a letter had to be procured from a clergyman testifying that I was of good moral character which made me feel that I was destined for Roedean or Eton, or some similar establishment, rather than the fo'c'sle of a Finnish four-masted steel barque.

Altogether Lubbock wrote about a dozen books and he edited the Log of the *Cutty Sark*. The majority of them were about the cargo and passenger carrying sailing ships of the mid-nineteenth and early twentieth centuries, about the men who sailed them, the life aboard and the passages they made. Among some of the varied cargoes they carried were opium, tea, coal, case oil, nitrate, grain, guano, a disagreeable cargo of seabird dung used as fertiliser, and grand pianos. And besides the fare paying passengers, emigrants and prospectors, some of them carried, imprisoned under hatches, (and on one occasion when a ship caught fire, they were roasted to death) wretched, enslaved Chinese coolies taken from their homeland to such far off places as Havana, or else to dig the guano from the ninety feet deep deposits of the stuff on the hellish Guanape Islands off the coast of Peru, eight degrees south of the Equator, from which few of them, as did the convicts who shared their labours, emerged alive.

All these books are classics of their kind, intended, as the author wrote in the preface to *The China Clippers*, for sailors. And in fact between the wars there were few British masters and mates who had served their time in sail, who did not have at least one of Lubbock's works on their cabin shelves. To some extent they *are* technical but one of the good things about Lubbock and his writing is that although he is an expert—he was quick to learn about sailing ships and their handling on his voyage home in 1899—he has the capacity of being able to make technical matters not only comprehensible to the landsman, but at the same time interesting.

A clipper, according to *The Shorter Oxford Dictionary*, as well as being "a vessel with sharp forward-raking bows and masts raking aft", is anyone or anything "which moves swiftly, or scuds along", and this is what a clipper was in both senses of the word. It was the Americans who invented this kind of ship which scuds along during the War of Independence. It was known as the Baltimore Clipper and its lines were borrowed from those of visiting French vessels which were far finer than their bluff, British counterparts. They were comparatively small, brigs and schooners mostly, few of them full-rigged in the proper sense with square sail on all three masts, and in the 1840s they competed with the British, who by this time were using sharper vessels themselves, in running opium from the Hooghly to the China coast, a trade which Lubbock deals with in greater detail in his book *The Opium Clippers*.

It was not until the 1850s that what are now recognised as archetypal clipper ships began to be built in any number on the shores of New England by such master shipwrights as Donald Mackay of East Boston. They were vessels of between 1500 and 2500 tons, built of softwood, square-rigged on all three masts and carrying immense and towering sail plans, intended to be driven hard with big crews in the North and South Atlantics and while beating around Cape Horn against the prevailing westerlies. They were given wonderful names, all of which became household ones to sailors: *Staghound*, *Flying Cloud*, *Lightning*, *Flying Fish* and *Sovereign of the Seas* were some of them.

In this book, *The China Clippers*, Lubbock is principally concerned with the birth, life and ultimate death of the ships, British and American, engaged in the China tea trade, perhaps the most beautiful ships ever to sail the seas. Built originally of either soft or hardwood, according to whether they were American or British, and later of hardwood with iron frames, of anything up to about 1000 tons and carrying great clouds of canvas—with all her flying kites set *Sir Lancelot* carried some 35,000 square feet of sail—the British ships could ghost along at five or six knots when a Mackay clipper, built to ride the biggest seas in the world off Cape Horn would barely have had steerage way. It was said of *Thermopylae*, the great rival of *Cutty Sark*, that she could sail seven knots while a man walked round her deck carrying a lighted candle. No wonder they were called the "China Birds". Not only were they good in light airs, they could beat dead to windward in a strong breeze, and with all sail set run 12–13 knots comfortably on a quartering breeze. On the other hand when running before the wind in heavy weather, many of them were very wet ships, scooping up the following seas over their sterns and inundating their decks while a big American clipper would be running with decks as dry as a bone.

Even the toughest sailors were susceptible to their beauty and their handling qualities. Of *Ariel*, a composite ship built of hardwood with iron frames on the Clyde in 1865, her captain wrote:

> "*Ariel* was a perfect beauty to every nautical man who saw her; in symmetrical grace and proportion of hull, spars, sails, rigging and finish, she satisfied the eye and

put all in love with her . . . It was a pleasure to coach her. Very light airs gave her headway and I could trust her like a live thing in all evolutions: in fact she could do anything short of speaking. *Ariel* often went 11 and 12 knots sharp on a bowline (sailing close-hauled into the wind), and in fair winds 14, 15 and 16 knots . . ."

Such an effect did *Thermopylae* have on the captain of a ship of the Royal Navy who endeavoured to race with her that he signalled as his ship dropped rapidly astern, "Goodbye. You are too much for us. You are the finest model of a ship I ever saw. It does my heart good to look at you."

The great years of the tea clippers were from about 1850 to 1873, with the American clippers ceasing to participate in the China trade after about 1855. The great races to reach the Channel and London with the new season's tea took place every year from 1859 to 1870 and Lubbock's book gives vivid descriptions of how the ships that were to race home used to assemble off Foochow in the Min river each April and May to load the tea brought down in sampans in matted chests; the ships a wonderful sight with their snow-white decks, snow-white because men had worked for days on end down on their knees holystoning them. In the American ships, which were hell to work in, because of the brutality of the officers who battered the men, sometimes to death, with handspikes and belaying pins, the decks were holystoned, then coated with coal tar, then holystoned again.

The crews of the British ships were almost all British compatriots and conditions on board were on the whole very humane. They were wonderful sailors. With a crew of only forty, which included master, officers, cooks, stewards, carpenters, sailmakers and so on, they could lift their anchors and make sail in twenty minutes. They could tack ship and have everything coiled down neatly afterwards in ten. They could deal with anything—a dismasting in a typhoon in the China Sea or off Ushant in the Channel. When *Cutty Sark* lost her rudder in the Indian Ocean, racing home from Shanghai against *Thermopylae* in 1872, they made a new jury rudder and she arrived in the Channel less than a week after her rival. Every AB carried a sail needle and palm and could repair a sail. They were very proud of their ships. In the Great Tea Race of 1866, the crews of *Serica* and *Fiery Cross* bet each other a month's wages that they would be first home.

Amongst the most masterly evocations in *The China Clippers* is the author's description of The Great Tea Race of 1866 in which five ships, *Fiery Cross*, *Ariel*, *Taeping*, *Serica* and *Taitsing* left the Min river at the end of May, each of them loaded with between 850,000 and 1,000,000 pounds of tea. *Fiery Cross* was the first to leave, followed by *Ariel*, *Taeping* and *Serica*, all three on the same tide in thick weather, then *Taitsing*, a day behind, all of them within two days of each other. He describes how they raced down through the Formosa Strait into the South China Sea which at that time was almost as infested with uncharted reefs and banks as it was with pirates, then through the Java Sea and out through the Sunda Strait between Java and Sumatra into the Indian Ocean where they set every stitch of sail they possessed, running for Mauritius 300 miles and more a day in the S.E. Trades, which they reached in the same order that they had left the Min river. They were still in the same order when they reached the longitude of the Cape of Good Hope, *Ariel* being forty-four days out from Foochow when she passed it; and the race culminates in *Ariel*, *Taeping* and *Serica* storming up the Channel in full sail, something that those who saw them would never forget, to take their pilots at Dungeness, then all three docking on the same tide in the London river, ninety-nine days from the Min, with *Fiery Cross* and *Taitsing* both coming in 101 days out, close behind them.

The Suez Canal was opened in 1869, the year the *Cutty Sark* was built, a year after *Thermopylae*. It was the end of an era. The last year in which there was any real racing was 1873, although the trade still continued. (*Thermopylae* made her last real passage from Foochow to London in 1881.) Both *Thermopylae* and *Cutty Sark* were lucky. They were big and strong and seaworthy, and they were just as happy employed in the Wool Trade and storming down to Australia in the Southern Indian Ocean as they had been ghosting along in the S.W. Monsoon in the China Sea.

Eric Newby 1984

PREFACE TO THE ORIGINAL EDITION

THIS book is an effort to preserve the records of the most perfect type of sailing ship at the very height of its development, and it has been written entirely for sailors and those who are interested in shipping. In it I have put down as simply as possible the personal history of certain ships and that in the plain language of the sea without any attempt to explain technical or seafaring terms for the benefit of the landsman. At the same time the reader must not expect to find highly-coloured descriptions of great sea adventures—the adventures are there right enough, but he who wishes to find them must search deep into the cold print of bald statements and read between the lines by the light of his own experience.

The material gathered together in this book has been culled from countless abstract log books, as well as from information supplied to me, not only by the men who sailed the ships but also by their owners, designers and builders.

Indeed I have to thank so many people for their help that a page of print would not contain their names, and I can only hope that this book may, perhaps, recall some pleasant sea memories and thus in some slight way recompense them for their kindness and trouble.

NOTE.—Where the word mile is used in this book, the sea-mile or 6080 feet is always meant, not the statute or land mile which is only 5280 feet.

CONTENTS

PART I.

CONTENTS

PART II.

CONTENTS

CONTENTS

APPENDIX

THE CHINA CLIPPERS

THE CHINA CLIPPERS

PART I

O fair she was to look on, as some spirit of the sea,
When she raced from China, homeward, with her freight of fragrant tea
And the shining swift bonito and the wide-winged albatross
Claimed kinship with the clipper beneath the Southern Cross.

Close-hauled with shortened canvas, swift and plunging she could sweep
Through the gale that rose to bar her wild pathway on the deep;
And before the gale blew over, half her drenched and driven crew,
To the tune of "Reuben Ranzo," hoisted topsail yards anew.

From the haven of the present she has cleared and slipped away,
Loaded deep and running free for the port of yesterday,
And the cargo that she carried, ah! it was not China tea.
She took with her all the glamour and romance of life at sea.

—K. TARDIF.

The Baltimore Clippers.

THE first ships that were ever built with speed instead of carrying capacity as the chief desideratum were the long, low, flush-decked Baltimore brigs and schooners, which by reason of their unusual sailing powers became celebrated the world over under the name of the "Baltimore Clippers".

These vessels dated from as far back as the American War of Independence, many of them were privateers, still more of them were slavers, whilst not a few ranged the Indies with the dreaded skull and crossbones flying from their signal halliards. These Yankee free-lances were wonderfully speedy in light airs and in turning to windward; and, carrying as they did large desperate crews and heavy armaments, often proved themselves more than a match for the tubby, overmasted sloops and brigs of the British Navy.

The Baltimore type had several very striking peculiarities

found in no other ships of the same date. It is supposed to have first originated in St. Michael's, Talbot County, where the art of shipbuilding had been handed down from father to son for generations.

Its chief features were great beam, placed far forward, giving a very fine run from a high bow with plenty of sheer to a low stern. Both stem, sternpost and masts were unusually raked, and it was this feature in the masts of a ship, together with a low freeboard, which, in the eyes of a stranger, gave immediate cause for anxiety and alarm, for any vessel described by the lookout as a" rakish looking craft" was at once suspected of being an ocean free-lance.

Other characteristics of the Baltimore clipper were long, easy water lines, with nothing concave about them like those of the later American clippers, great dead rise at the midship section, and an unusual number of flying kites in addition to her working sails, which were, of course, like all American sails of gleaming cotton duck.

For some time the word clipper was only applied to Baltimore ships, but gradually as fast ships began to be built in other ports for trades in which speed was of great importance, these also were called clippers, though none of these later clippers in the least resembled the Baltimore schooners, with the exception perhaps of some of the American opium clippers, which certainly did show many of the characteristics of the Baltimore model.

The Opium Clippers.

From early in the nineteenth century to the late seventies three distinct types of fast sailing ships or clippers were employed in the China trade. These were the opium clippers, the American tea clippers and the British tea clippers. The first dated from about 1830 to 1850, the second from 1846 to 1860, and the last from 1850 to 1875.

Before turning to the great China tea trade and the wonderful ships it produced, I will first try and convey some idea of the opium trade and its clippers.

From the very start the importation of opium into China was entirely against the decrees and wishes of its rulers, who knew only too well the harm done by the drug to all who fall under its influence. And those enterprising British, American and Parsee firms who engaged in the opium traffic were nothing more nor less than smugglers, smugglers indeed who showed greater daring and finer seamanship and made bigger profits than any the world had previously known.

I have no intention here of touching upon the morality of the trade, suffice it to say that there is more than a little truth in the yarn that misfortune and unhappiness always followed those who had made their fortunes in opium.

I may safely say, however, that the opium traffic did this much good: it produced some very fine ships and it trained an incomparable race of seamen.

For such a hazardous trade swift keels were a necessity. At the close of the Government opium sales in India, the rich dark cakes of Patna and the shapely balls of Benares, which composed the new crop, were shipped into small clippers, which were specially built for the trade. These vessels had to make the passage round to China under racing canvas at all seasons and weathers, and especially during the strength of the N.E. monsoon, when they had to thrash their way to the Chinese Ladrones against a heavy head sea and strong current, either in the open or by the Palawan Passage.

At Linton or Macao they generally transhipped the precious drug into receiving ships, which were as a rule old Indiamen transformed into floating warehouses. These receiving ships* were also stationary gunboats with large fighting crews in addition to staffs of schroffs and clerks who attended to the sales and other business. A third class of smaller vessel fed the opium from these receiving ships to other clippers stationed along the coast between Hainan and Woosung.

---

* In 1850 the receiving ships at Namoa were:—*Anonyma*, brig (formerly Colonel Greville's yacht), Jardine & Co., and *Hong Kong*, barque, Dent & Co.

At the six islands near Amoy there were:—*Lord Amherst*, barque, Dent & Co.; *Pathfinder*, barque, Jardine & Co.; *Royalist*, schooner (formerly Rajah Brooke's yacht), Syme, Minn & Co.

Besides these there were vessels stationed at Foochow, Woosung, Chinchew and in the Cap-Sing Moon Passage, near Hong Kong, both English and American.

This fourth class, which were usually the pick of the lot, had the duty of carrying the opium to places where no treaties or agents existed. Indeed, theirs was the most arduous, if most exciting task of all. They had to meet the Chinese opium smugglers in lonely creeks which had never been surveyed, knowing that these same smugglers would be only too ready to capture their clipper if given the chance or to loot her and murder her crew if she stranded. They had to circumvent the wiles of hostile mandarins, defend themselves not only against war junks but fleets of Chinese pirates; weather the dreaded typhoons and, if damaged, refit themselves at sea; open up new trade with far away and unknown ports; survey new coasts and harbours, carry mails and despatches and even negotiate treaties.

In such a perilous trade the officers had to be carefully picked. In the British opium clippers many of them were ex-naval men, but there was evidently as great a competition for the appointments as there was for the service of the old "John Company".

An officer, who served aboard the *Falcon*, most famous of all the clippers, writes:—

"The officers were for the most part the younger sons of good families at home, who had to use every effort and wait long to fill a vacant appointment, which was very difficult to obtain, as applicants had to undergo the severest tests of fitness, both mental and physical. Some acquaintance with nautical astronomy and the physical sciences, with a taste for Eastern languages and a tongue and turn for Eastern colloquials; approved physique, steadiness and courage; reliability of temper, with the higher moral gifts of coolness and patience under trial and provocation—all these were essentials. And it may be remarked that among the officers were many sons of clergymen, who, after a period of active service afloat, would retire to succeed ultimately to their father's livings or to practise at the bar, not a few finding their way into Parliament".

The pay was enough to make a present-day sailor's mouth water. The captains, if they succeeded in avoiding capture, very soon made fortunes and retired. Each officer was allowed his own Chinese boy or body servant, whilst the captain had his butler and two boys.

The clippers carried double crews, composed of all nationalities, but amongst whom were always to be found a sprinkling of deserters from the Royal Navy, drawn by the lure of high pay and promise of excitement. Discipline had, of course, to be very strict; gun and cutlass drill formed a regular part of the routine, whilst smartness in sail handling was a matter of *esprit de corps*.

The chief firms engaged in the trade were the British firms of Jardine, Matheson & Co. and Dent & Co., whose establishments at Hong Kong were on the most magnificent scale. Russell & Co. represented American interests, whilst the Parsee family of Bonajee represented Indian.

The opium clippers were all small vessels, mostly brigs and schooners, with a few barques and only one ship, the celebrated *Falcon*; but size was not required, for their only cargo was opium and silver specie.

An opium cargo rarely consisted of more than two or three hundred chests, which were exchanged for silver, usually in the shape of Mexican dollars but sometimes in bars. Occasionally, at out-of-the-way places, antique ware in the shape of vases and ornaments of gold or silver were accepted for opium, and in this way, I fear, many a precious work of art found its way into the melting pot.

The piratical lorchas were always on the lookout for a becalmed opium clipper, for whether upward bound with the drug or downward bound with specie she was always certain to be a rich prize. But in order to overcome this danger all opium clippers were supplied with 40-foot sweeps which were run out of the gun ports and, with six men to each, would send a schooner along 3 to 4 knots with ease.

There was one still greater danger in Chinese seas than the pirate and that was the typhoon. Captain Anderson gives a very vivid description of a typhoon which he experienced when third officer of the opium schooner Eamont.

"The *Eamont* was bound from Hong Kong to Amoy. At 10 p.m. the wind was abeam, blowing off the land, the schooner going 12 knots with a smooth sea and not a cloud in the sky. At 11 p.m. the wind became gusty, dark clouds began to form ahead, and the barometer had fallen an inch since four bells. The captain was called. He at once kept the ship off the land 3 points, and had all hands called to shorten sail. Whilst the topgallant sail was being taken in and the yard sent down on deck, the wind was increasing in noisy gusts. A bo's'n's mate and eight men were unable to secure the gaff topsail, which had a 28-foot yard, and filling like a balloon, blew clean away. And before the topsail could be put in its gaskets, it flew away in pieces not more than a yard or two in size.

"1 a.m.—The ship was scudding under stay foresail and half the fore trysail, the lower half having blown away whilst the sail was being lowered. The sea was now a mass of seething foam, the rain fell in solid spiral columns and the wind was blowing so hard in the squalls that the ship lay down under it with her hatches in the water. Several attempts were made to bring her to, but each time she was knocked flat with her cross-trees barely clear of the smother to leeward. The barometer was 27·80. The two lee boats and davits were torn away, whilst the weather boats were smashed to pieces against the shoulders of the davits by the mere force of the wind, which had now the true cyclone howl.

"4 a.m.—The barometer was 27·50. The Chinese cooks and stewards lay about in the wreckage of the saloon, but the third officer and two quartermasters managed to serve out a stiff glass of grog and a snack of grub all round.

"The wind now roared heavier than the crash of a battery of heavy artillery and nothing was to be seen from the schooner but a curtain of seething foam. The *Eamont* scudded three times round the compass in an hour.

"Daylight.—The wind made it impossible to go aloft as one could not back one's feet out of the ratlines. The four lee guns chafed through their breechings and went to fit a frigate for Davy Jones.

"6 a.m.—The squalls came up every quarter of an hour.

"8 a.m.—The squalls came up every ten minutes.

"10 a.m.—The squalls came up every five minutes.

"Barometer 27·30. The carpenter and hands were stationed at the masts and weather rigging with axes, ready to cut away at the order. The square foresail which hung up and down from the slings of the foreyard, and besides having four pairs of brails was secured with extra lashings, blew adrift, filled like a balloon and threatened to take the foremast out of her. With great difficulty it was cut away from the yard.

"11 a.m.—The squalls coming up in such rapid succession that it seemed to be one long dismal howl. A tremendous squall laid her flat on her broadside with the cross-trees in the water. The captain shouted 'cut' through his trumpet to the men with axes, but a second later roared 'Hold on, all'. In that moment the wind and rain ceased as if by magic, the ship stood up becalmed and began to jump about in a curious turmoil of sea, which was running in every direction. Not a breath of air stirred and a dense misty cloud hung all round to the distance of about 2 miles. The *Eamont* was caught in the calm centre of the cyclone.

"11.15 a.m.—A new fore staysail and inner jib were got up and bent.

"11.30 a.m.—The hands were starting to sway up a new topsail when the wind came again in a deafening roar, and striking the schooner on the starboard beam (the same side as before, luckily) sent her on her beam ends, the new fore staysail and jib parting their sheets with reports like heavy guns and flying away into space.

Noon.—Barometer 27·60. Wind south on starboard quarter, ship's head N.E., ship going 12 knots. Typhoon, moving to westward, enabled ship to be brought up to E.N.E.

"3 p.m.—Wore ship and set course N.W. for Amoy.

"6 p.m.—Set topsail, new fore trysail, staysail and jibs.

"Sunset.—Land sighted about 18 miles on port beam.

"Since passing through the centre of the typhoon, the *Eamont* had scudded N.E. about 40 miles, then gone course 40 miles, so that when madly scudding round the centre in the morning, she could never have been more than 12 miles off the land".

The year before she encountered this typhoon, the *Eamont* rode one out off Swatow, hanging to one chain with 150 fathoms out and a second anchor backed on it at 60 fathoms. They had to cut her masts away, but they hung the wreckage astern and four days later stepped the shortened masts and, setting the sails double-reefed, fetched Amoy in three days. I will give one more instance of an opium clipper in a typhoon. In October, 1848, the American clipper-brig *Antelope*, Captain Watkins, ran into a typhoon when bound from Shanghai to Hong Kong, and the following is her captain's letter to his owners:—

"I regret to inform you that we have been totally dismasted. We left Woosung on the 5th October in fine weather. On the 6th, blowing fresh from the east, the brig going large 12 knots; at 6 o'clock we double-reefed the topsails and furled the mainsail; at midnight took in close reefs, the barometer at 29·35; at 4 a.m. furled the fore topsail and sent down topgallant yards, the wind and sea increasing, washed away both quarter boats and part of the hammock rails. I was compelled to keep running so as to get sea-room to heave to; the wind shifted to north and blew a perfect hurricane; barometer 29·30. At 3 p.m. the foresail blew away, a dangerous cross sea running. At 4.30 p.m. prepared to heave to, every aperture was carefully battened down, and all spare spars and the launch carefully secured. At 5 p.m. rounded to. At 5.20 a furious squall and she went over on her beam ends, the helm being no longer of use, she fell off in the trough of the sea. I knew that if she remained so she must go down, and I ordered the masts cut away, and by 7 a.m. we were clear of the wreck, barometer 29·10 and the sea making a clean breach over her. On the 8th, at 7 a.m., let go the stream anchor and the kedge in 21 fathoms of water in order to try and get her head to the wind. We soon drifted into 12 fathoms on the Formosa Banks, still blowing heavy and the sea worse than yesterday. At 3 o'clock drifted into 19 fathoms. At 8 p.m. the stream and kedge were lost and she fell off, bringing the wind on the starboard quarter. I dared not let go a bower anchor, for in the event of a shift of wind I might want the anchors. At 9 p.m. still blowing a heavy gale. At midnight the wind came round to S.E., barometer 29·40.

"On the 10th, at 4 a.m., wind came out of the S.W. and moderated fast, barometer 29·80. At daylight commenced to rig jury masts. At noon, calm, and by 9 p.m. we had the brig under snug working canvas. I am satisfied that if the *Antelope* had not been one of the best built vessels afloat she must have gone to Davy Jones' locker. When we left Shanghai she was a perfect picture; we arrived on the 13th and since then seven others have arrived in a similar condition, and some will never be heard of".

After this the *Antelope* was re-rigged as a barque, and was never quite the same vessel.

I will now turn to the ships themselves.

The "Falcon".

The celebrated *Falcon* was the only shiprigged opium clipper that I know of. This vessel had a very interesting career. She was built by List, of Wootton Creek, near Cowes, and was launched on 10th June, 1824, her owner being the Earl of Yarborough, commodore of the R.Y.S. She was built regardless of expense and in appearance resembled a 20-gun corvette. The Royal Yacht Club had been founded with more serious objects than that of summer yachting inside the Isle of Wight, and at this date possessed several yachts fitted as men of war.

It is related of Lord Yarborough that he paid his yachting crew extra wages on consideration that they conformed to the regulations of the Royal Navy. One of these regulations justified the free use of the cat o' nine tails, and before leaving Plymouth for a Mediterranean cruise all hands on the *Falcon* signed a paper setting forth the merits of a sound flogging and their willingness to undergo it, if needful, for the preservation of discipline.

The *Falcon* measured 351 tons, and while owned by Lord Yarborough mounted a broadside of eleven guns. On 20th October, 1827, she was through the thickest part of the action of Navarino, her sporting owner flying his flag as Admiral of the Isle of Wight at the main and his commodore's burgee at the fore. For some years Lord Yarborough regularly joined the cruise of the experimental squadron, whose chief object was to test the speed of new ships and help along the designing of fast and seaworthy men-of-war. Thus the *Falcon* tried her paces with the last of our wooden men-of-war, and proved herself a difficult vessel to pass. In 1835 Lord Yarborough, badly injured by being thrown across a sea chest in a gale of wind, and further disabled by an attack of influenza, decided to sell his famous vessel. She was bought by a London firm for £5500. They fitted her with two 24-horse-power engines and sent her out under sail to Calcutta, in the hopes that the Government would buy her for use in the Burmese war.

The *Falcon* sailed from Cowes in 1836, on the day that the King's Cup was being raced for, and it was with great regret that the members of the Royal Yacht Club watched the pride of the squadron sail away.

On her arrival at Calcutta, she was bought by Jardine, Matheson & Co. They removed the engines and fitted her in the most thorough and expensive manner for the opium trade.

I will now give a long quotation, with apologies to the *Yachtsman* from which it was taken. Written by an officer aboard the *Falcon*, the following not only gives one a valuable picture of the ship herself but a still more valuable one of the splendid seamen who served aboard her:—

"With a bow round and full above the waterline, she was as sharp as a wedge in her entrance below. Her midship sections gave her a long flat floor, whence commenced a clean run aft, that, with her form of entrance, minimised resistance and displacement to a marvellous degree. Her breadth of beam over this long floor enabled her to stand up under a more than ordinary press of canvas, while it afforded quarters for a small battery of guns, including a long brass piece amidships and some pivot and swivels over bows, counter and quarter, that made her a wholesome terror to the swarming fleets of pirates, which then infested the Hokkeen Coasts.

"In all cases of bad weather, the heaviest of these were run in and well secured, indeed there were times and occasions when the whole armament was dismounted and put under hatches, so that nothing should encumber the spacious white flush deck beyond the neat coils of running gear placed in tubs made for the occasion.

"The *Falcon* was a full-rigged ship, heavily but beautifully masted, as to rake and proportion. Her yards and spars were of dimensions equal to a ship of, perhaps, twice the size in actual carrying power in the ordinary Mercantile Marine. These were beautifully fashioned and finished—not in the tapered and pointed style affected by traders within the tropics, nor of the dilettanti in the summer seas at home, but in a style that savoured rather of massive strength and utility. There was no skysail or moonsail or flagstaff extensions. Our masts seldom went more than a few inches beyond the rigging that supported them.

"In summer-like weather we sent up topgallant and royal masts in one, but during the strength of the monsoons and in all passages to the northward—and we sometimes went very far north—short topgallant masts were fidded. We trusted more to spread than to hoist; and in going free the show of canvas upon our square yards, further extended by lengthy stunsail booms—in the rigging out of which our topmen had few compeers—would leave an observer in no doubt of the immense pressure under which the comparatively slight and beautiful fabric trembled and vibrated in its headlong career.

"Our spars from deck to truck were, or had been, modelled by rough and ready artists, in the persons of our carpenter and his mates, who had sometimes more than they could do to supply our frequent losses. Famous among us as he was—and as he deserved to be—our carpenter yielded to the superior art of our sail maker. Much as they did to enhance each others' merits, the sail maker bore the palm. No academician ever draped a classic figure with more consummate taste and art than that with which our sailmaker draped the *Falcon.* Nothing in still life could be more picturesque than the *Falcon's* sails, which, unfurled at anchor or in a calm or other condition of repose, fell in full, heavy graceful folds from her yards and booms. Nothing could convey so strikingly the triumph of art, when the same sails were filled and trimmed—full and by—in the first case presenting a cloud of swelling segments, pressing forward as if in spirited and living rivalry; in the second case held like boards by sheet, tack, brace and bowline, the rounded luff and foot leaving no rift twixt spar and canvas; in both cases gladdening and satisfying a seaman's heart and eye.

"At daylight every morning there was a general resetting of sails, a repointing of yards, and a 'freshening of the nip' in every sheave. Many watchful eyes and ready hands were on the alert, upon these and similar occasions, to make the slightest change of wind, whether in force or in direction, available to add a knot to the coming day's work.

"Long as I have lingered in my description of the carpenter and sailmaker, I cannot honestly proceed further without a word in praise of another deserving petty officer, the boatswain, who, with a couple of mates and four quartermasters, had come out from England in the *Falcon*.

"The boatswain had been a foreman rigger in one of the great commercial docks at home, where his daily practice for many years had familiarised him with every description of craft of every possible rig, and with fittings and refittings to suit almost every taste. He was a master of his craft, and was as intelligent as he was expert. His leading peculiarity was his faculty of teaching and of communicating to others not only the mode, but the philosophy and the spirit and beauty of his own gifts, so that the three years that we had spent together in the *Falcon* had made us all riggers.

"All able seamen, by mere use of the term, profess to be competent to 'hand, reef and steer and heave the lead'; but there wasn't a seaman on board the *Falcon*, who couldn't—besides these requisites—turn in a dead-eye, gammon a bow-sprit, fish a broken spar, rig a purchase of any given power, knot, point, splice, parcel and serve, spinning his own yarn or lines, of such length and dimensions as could be adapted to the power of our winch and rope-walk. With such a crew the state of our rigging, stays, backstays, standing and running gear and fittings may be accepted as most perfect and complete. As an instance of the capacity of the *Falcon's* crew, I may state that we have stretched, cut and fitted a set of coir lower rigging on our own decks at sea; and at sea we have placed it over the naked lower mastheads, and set it up, one mast at a time, completing the whole work in three days. And at sea we have constructed a raft of spare spars, and transferred to it our guns, stores and much of our ballast. And to the raft so loaded, we have hove down the ship and repaired and cleaned the copper from keel to bends in two days. Our crew was a large one, I admit, sufficient in number to make three strong watches, either of which could reef the three topsails together. But I would remind my nautical reader that no number—no mere number—of unskilled or undisciplined men could have been trusted to perform tasks so onerous as those I have described. If I am asked why such feats should have been performed at all, or even attempted, I may say that the *Falcon* was always at sea; that her cruising grounds were over a long extent of coast that, in those days, swarmed with pirates; that it afforded no place of shelter where strangers could safely enter; that the whole coast was *terra incognita*, except to the lawless, rapacious natives and to the few like ourselves who had spent many years in its navigation in all weathers. There were no steamers in those days, the first that ever visited China was, I believe, the *Jamesina* from Bombay.* And the first complete survey of the coast was made ten years after this date by Lieut. Collinson, R.N. He did the most of his work in an old Calcutta pilot brig, the *Plover*, and did it well in the face of great perils, hardships and obstacles, writing sailing directions and calculating and compiling marine statistics and astronomical data of surpassing accuracy and value.

"I hesitate to touch upon the secondary furnishings and fittings of the *Falcon*, but I remember that they were unusually elegant and substantial and costly; that where metal was employed it was mostly of copper or brass, even to the belaying pins; and that toprails, stanchions, skylights, and coamings were of mahogany, whilst the accommodation below for officers and crew was extravagantly luxurious.

"It will better become me to speak of her higher and greater qualities, qualities which made her an object of pride and real affection, as of the tenderest care, of her crew, officers and men. She was easy, handy, and smart in every evolution. She swam like a duck, and steered like a fish. She was fast, yet dry; lively,

---

* When the *Jamesina* from Bombay entered the Canton river, and had taken her pilot on board at Lintin, as was the old custom, and was being propelled against wind and tide, the pilot gave occasional directions to the helmsman as if the ship had been sailing before the wind and with the tide in her favour and carefully abstained from any show of surprise or curiosity. On the captain drawing his attention to a fact which must have inwardly astonished him, he coolly replied that the mode of propulsion was no secret in some parts of the empire, where it had once prevailed, but had then *fallen into disuse*.

yet stiff. Sensitive and responsive to every yard of canvas that could be judiciously spread, as to every touch of the braces, tacks and sheets, and to every spoke of the wheel.

"It was in the *Falcon* that I learnt to comprehend and to adopt a singular belief that prevails among seamen; and it was in her, and by her, that I was first touched by that strange sympathy which is created by a favourite ship upon the minds of an appreciative crew. If the *Falcon* had been a living being that sympathy could scarcely have been greater. She would resent every neglect in her handling, and rebel at once against any overpressure or any tampering with her trim, so that our common expressions—expressions that could have no meaning to a landsman—that she was complaining or sulking or huffed or offended seemed to us to be rightfully applied.

"One felt proud to watch her dealing with opposing forces so persistently and so gallantly. We had been afloat in her for upwards of three years with few losses and fewer changes than could have been expected in so large a crew; and, having watched and studied her pretty ways for so long a period, we had acquired readiness and skill in her management, and had learnt to look upon her as a thing to be loved and petted' 'She can do everything but speak' was a common remark among the crew".

"Sylph" and Other Pioneer Opium Clippers.

One of the most celebrated of the earlier opium clippers was the *Sylph*, a 305-ton schooner, built at Calcutta in 1831 for the Banajee family. She is supposed to have run from the Sandheads to Macao in sixteen days. During the Chinese War of 1841, the Commander-in-Chief tried hard to buy her for the Navy, but her owners refused, preferring rather to keep her carrying opium at such a time, when every successful cargo run meant a huge profit.

The *Sylph* and another well-known clipper, the *Cowasjee Family*, were fitted out with extra guns and full European crews during the war, and were joined by the *Lady Hayes*, belonging to Jardine, Matheson & Co., the three ships sailing in company. Among the islands they were surrounded by Chinese war junks, and a fierce battle ensued. But Captains Vice and Wallace, of the

Sylph and *Cowasjee Family*, were two of the most experienced captains in the trade, celebrated for their daring and success in dealing with pirates, and the war junks suffered a severe defeat, many of them being sunk; after which the opium clippers had no more trouble.

Amongst the contemporaries of the *Sylph* and *Cowasjee Family* the best-known clippers were the *Jamesina* 382 tons (formerly H.M.S. gun brig *Curlew*); the *Red Rover*, a barque of 255 tons; the schooners *Rustomjee Cowasjee*, *Waterwitch*, *Mermaid*, and *Black Joke* (formerly a slaver), the *Ernaad* and *Ternate*, old East India Company cruisers; the *Forth*, *Pearl*, *Corsair*, *Venus*, and *Royal Exchange*.

The exciting lives of these vessels would make good reading if they could only be exhumed from the old log books, many of which could doubtless be found locked away in the carefully preserved sea chests of dead and gone sea captains.

American Opium Clippers.

The first American opium clipper is supposed to have been the schooner *Anglona*, of 90 tons, built by Brown & Bell, and sent out in 1841 for Russell & Co. She was followed by the *Ariel*, of 90 tons, *Mazeppa*, of 175 tons, and the *Zephyr*, of 150 tons, in the following year.

The *Ariel* was built by Sprague & James, of Medford, to the designs of Joseph Lee. She capsized on her trial trip in Boston harbour, and sank in 7 fathoms. On being raised, 8 feet were cut off her masts, the foremast was put further aft, and a false keel, 8 inches deep, added.

On her arrival out in China she raced round Lintin from Macao Roads against the *Anglona*. The latter had the best of it beating out, but in the run back *Ariel* gained 17 minutes, and won the sporting wager of 1000 dollars.

The *Zephyr* was built by Samuel Hall, of East Boston, on the American pilot boat model. Though lightly built, like most American ships, of American oak and elm, she was beautifully modelled and very heavily sparred; and she carried an armament of four brass 18-pounders to a broadside, a brass 18-pounder pivot gun on the foc's'le, also of brass, and, in her prime, an Armstrong 68-pounder on a pivot between the masts.

Her captain, T. M. Johnson, wrote home in 1843:—"The *Zephyr* is now in the Taypa with loss of main boom, fore-gaff, and jib. I was caught off the Grand Ladrone in a gale at north-east. She did well till the jib was lost, and we beat from the Ladrone to here under double-reefed sails and storm jib quicker than anything in China could have done it. None of the vessels I have met could beat her. On the wind I do believe there is not anything that can beat her. When in ballast she is as dry as any of them. In smooth water, on the wind, her equal is not to be found in China or elsewhere in my opinion. Properly ballasted she is the easiest vessel I was ever aboard of."

Two other celebrated American clippers were the brig *Antelope*, of 350 tons, whose experience of a cyclone I have already related, and the barque *Coquette*, of 450 tons.

The *Antelope* made the passage from Calcutta to Singapore in twelve days in 1844, and the *Coquette* did the same time in 1845, beating such well-known clippers as *Red Rover*, *Rob Roy*, *Mischief*, *Don Juan*, and *Kelpie* and reaching Hong Kong in thirty days. The *Antelope* had the reputation of being one of the few square-rigged vessels of that date which could beat through the Formosa Channel against the strength of the N.E. monsoon.

The following account of her first opium run I have taken from an American book of voyages, which, though of extraordinary interest, has been long forgotten and out of print:—

"There was just then lying in the River Hooghly one of the prettiest little craft that was ever in the opium business. She was called the *Antelope*, and had only come out from Boston six months before. With her low, black hull, tall rakish masts, and square yards, she was a regular beauty, just such a vessel as it does an old tar's heart good to set eyes on—though, for the matter of comfort, keep me out of them, for what with their scrubbing and scouring in port and their carrying on sail at sea to make a good passage and half drowning the crew, there's very little peace aboard them. We went aboard to take a look at the beauty, and before we left her had shipped for the voyage. The captain was a lank West

Indian, a nervous creature, who looked as though he never was quiet for a moment, even in his sleep, and we afterwards found out he didn't belie his looks. After taking a cruise around Calcutta for a couple of days we went on board, bag and hammock (for no chests were allowed in the forecastle). Our pay was to be 80 rupees per month with half a month's advance.

"The vessel was well armed, having two guns on a side besides a Long Tom amidships. Boarding pikes were arranged in great plenty on a rack around the main mast, and the large arms chest on the quarter-deck was well supplied with pistols and cutlasses We were fully prepared for a brush with the rascally Chinese, and determined not to be put out of our course by one or two mandarin boats.

"We sailed up the river some miles to take in our chests of opium, and, having them safely stowed under hatches, proceeded to sea. With a steady wind we were soon outside the Sandheads. The pilot left us, and we crowded on all sail, with favouring breezes for the Straits of Malacca. If ever a vessel had canvas piled on her it was the *Antelope*. Our topsails were fully large enough for a vessel of double her tonnage. We carried about all the flying kites that a vessel of her rig has room for. Skysails, royal studding sails, jib-o-jib, staysails, alow and aloft, and even water sails, and savealls to fit beneath the foot of the topsails.

"She steered like a top, but our nervous skipper, who was not for a moment day or night at rest, but ever driving the vessel, had one of those compasses in the binnacle, the bottom of which, being out, shows in the cabin just how the ship's head is at any moment. Under this compass, on the transom, the old man used to lay himself down, when he pretended to sleep (for we never believed that he really slept a wink); and the vessel could not deviate a quarter of a point off her course, or while we were on the wind the royals could not lift in the least, before he was upon the helmsman, cursing and swearing like a trooper, and making as much fuss as though she had yawed a point each way. It was the season of the south-west monsoon, and, of course, we had nearly a head wind down through the Malacca Strait. But our little craft could go to windward, making a long tack and a short one nearly as fast as many an old cotton tub can go before the wind.

"Our crew consisted of seventeen men, all stout able fellows. There were no boys to handle the light sails, and it was sometimes neck-breaking work to shin up the tall royal mast when skysails were to be furled or royal studd'nsail gear rove. We had but little to do on board; to mend a few sails and steer the vessel was the sum total of our duty. On board these ships the men are wanted mainly to work ship expeditiously and to defend her against the attacks of the Chinese officers, whose duty, but ill fulfilled, it was to prevent the smuggling of opium into the country.*

"Once past Singapore we had a fair wind, and, with all studding sails set, made a straight wake for the mouth of the Canton River. As we neared the Chinese coast preparations were made for repelling any possible attacks. Cutlasses were placed on the quarter-deck ready for use, pistols loaded, and boarded nettings rigged to trice up between the rigging some 10 feet above the rail, thus materially obstructing any attempts to board the vessel when they were triced up.

"It did not take our little clipper many days to cross the China Sea. We had passed the Ass's Ears, the first landfall for China-bound vessels, and were just among the Ladrone Islands, which are the great stronghold of Chinese pirates, when we beheld starting out from under the land two of the long mandarin boats. They appeared to know our craft or to suspect her business, for they steered straight towards us. With the immense force they have at the oars it did not take them long to get within gun-shot range, which was no sooner the case than our skipper, taking good aim, let fly a shot from Long Tom in their midst. This evidence of our readiness for them took them all aback, and after consulting together for a little, they showed themselves to be possessed of the better part of valour—prudence—by retreating to their lurking place behind the land. We knew that so long as we were in clear water and had a good breeze, there was but little to be feared from them. The only danger was in case we should be becalmed when we got under the lee of the land, as they would be keeping a constant watch upon us and in such case would make a desperate rush upon

* This was before the first Chinese War, and when the mandarins were still making feeble efforts to stop the opium coming into the country.

us and perhaps capture us by mere superiority of numbers. As may be imagined, we were all determined to defend ourselves to the last, even the black cook kept his largest boiler constantly on the galley stove, filled with boiling water, wherewith to give the rascals a warm salute should they endeavour to board.

"What we feared shortly came to pass. In less than two hours after we had seen the boats, we lay becalmed under the land. The little vessel was perfectly unmanageable, drifting at the mercy of the current. Had we been far enough inshore we should have anchored. As it was, we could neither anchor, nor could we manage the vessel to turn her broadside towards the enemy.

"It was not long before they made their appearance. They had in the meantime obtained reinforcements, and four large boats, containing from sixty to a hundred men each, now shot out from under the land, and came toward us with rapid sweeps. We did not wait for them to come to close quarters, but sent some shots at them from Long Tom. These, however, did not deter them. The calm had given them courage, and after discharging their swivels at us, with the hope of crippling the vessel by hitting some of our tophamper, they rushed to the onslaught.

"We now rapidly triced up our boarding nettings, and lying down under shelter of the lower rail awaited the attack. The boarding nettings they were evidently unprepared for, as at the sight of them they made a short halt. This the old man took advantage of, and taking good aim, let drive Long Tom at them, and luckily this time with good effect, knocking a hole in one of the boats, and evidently wounding some of her crew. Taking this as a signal to advance, and leaving the disabled boat to shift for itself, the remaining three now rapidly advanced to board Taking advantage of the unmanageableness of our vessel, they came down immediately ahead to board us over the bow, a position where, they well knew they were secure from the shot of our light guns, which could only be fired from the broadside.

"Cocking our pistols, and laying the boarding pikes down at our sides ready for instant use, we waited for them. Directly twenty or thirty leaped upon the low bowsprit, some rushing to the nettings with knives to cut an entrance. We took deliberate aim and fired, about a dozen falling back into the boats as the result of our first and only shot. Dropping the firearms we now took to the pikes and rushed to the bow. Here the battle was for some minutes pretty fierce, and a rent having been made in the boarding net the Chinamen rushed to it like tigers, but as fast as they came in they were piked and driven back.

"Meantime one of the boats had silently dropped alongside, and ere we were aware of it, her crew were boarding us in the rear. But here the doctor (the cook) was prepared for them, and the first that showed their heads above the rail received half a bucket full of scalding water in their faces, which sent them back to their boat howling with pain.

"That's it, doctor, give it to them," shouted the old man, who seemed to be quite in his element, and he rushed down off the poop, whither he had gone for a moment to survey the contest, and taking a bucketful of the boiling water forward threw it in among the Chinamen, who were still obstinately contesting the possession of the bow. With a howl of mixed pain and surprise they retreated, and we succeeded in fairly driving them back into the boats.

"A portion of us had before this gone to the assistance of the cook and had succeeded in keeping them at bay aft. To tell the truth, the hot water frightened them more than anything else, and the boat's crew alongside required all the urging of their mandarin officer to make them charge at all. Luckily, at this moment, a squall which had been some time rising, broke upon us, and the brig began to forge ahead through the water. With a shout of victory we made a final rush at our assailants and driving them back into their boats, cut them adrift, giving the one alongside a parting salute of half a dozen shots in her botton, thrown in by hand. Making the best of our way from the scene of action, we steered towards Lintin Bay, where we were so fortunate as to meet a little fleet of opium boats, who quickly relieved us of our cargo, and we were no further molested by the mandarins, who had probably gotten a surfeit of the fighting".

On his next passage Captain Watkins was again attacked. This time he made a regular example of the mandarin boats, and after running down two of them, and drowning their crews, sailed into Macao Roads with a Chinaman hanging at each yardarm as a warning.

These fights, of course, occurred before any receiving ships were stationed near Lintin Island, and when the opium clippers had to tranship the drug into Chinese smuggling boats. And it was only on rare occasions that cargoes were run without fighting with the authorities in the shape of the Mandarins.

The last of the American opium clippers were the two 300-ton schooners *Minna* and *Brenda*, built by George Raynes, of Portsmouth, New Hampshire, in 1851.

Later British Opium Clippers.

As far back as the fifties steamers had begun to contest the opium trade with the clippers, and another ten years saw the end of these famous little ships.

Of the later British clippers the most notable were the schooners *Torrington*, *Eamont*, and *Wild Dayrell* and the brig *Lanrick*. The *Torrington* is interesting as being one of the first of Alexander Hall's clippers, having his famous Aberdeen bow. She was built in 1845 for Jardine, Matheson & Co., and proved a great success as a sailer.

The *Eamont* and *Wild Dayrell* were both built by White, of Cowes. The *Eamont* was constructed of teak and mahogany, measured a little over 200 tons and had a mainbroom 110 feet long. She was a very powerful vessel and carried 200 tons of iron kentledge fitted into her as a stationary ballast.

She was armed with four 18-pounders a side and two pivot guns, like the *Zephyr*, the two vessels belonging to Dent & Co. During the Taeping Rebellion the *Eamont* ran up to the threatened city of Ningpo, passing right through the battle of Chinhae, which was being waged not only on the banks but in the river itself. At Ningpo she found the *Zephyr*. The two schooners loaded up with fugitive celestials, and raced each other back to Woosung. But in the smooth sheltered water of the river, and with a fresh whole sail breeze, the *Zephyr* was more than a match for the more strongly built Cowes schooner. However, the *Eamont* had her revenge in weather more to her choice. The two vessels met this time in half a gale of wind with a heavy sea running, and the *Eamont* sailed right dead to windward of the *Zephyr*, and left her out of sight in twelve hours.

The *Eamont* was sent on some very dangerous trips. She was one of the first vessels to open up a trade with Formosa, and made the first survey of the port of Taku, which she entered by bumping over the reef in spite of a tremendous surf beating upon it at the time, a most daring performance. And in her efforts to trade with the savage Formosans she had to withstand the treacherous attack of hundreds of armed natives tight on top of a typhoon, which she succeeded in riding out on her moorings. But the Captain of the *Eamont* was a famous fighting man, as the Chinese pirates knew to their cost. In his constant encounters with piratical lorchas Captain Gulliver made use of a drag sail, with which he would suddenly deaden the way of his schooner, and so out-manoeuvre these "Invincibles," as they called themselves.

The *Eamont* was also employed in the negotiations for the first commercial treaty with Japan. On this occasion she ran into Nagasaki and quietly dropped anchor, in spite of the fact that opposition to the proposed commercial treaty was very strong at the time. On the following morning 150 boat-loads of Japs attempted to tow her to sea, being evidently ignorant of an anchor's *raison d'etre*. But though they attempted several similar methods to get rid of her they refrained from any armed attack, and, eventually, her mission was completely successful. This was in 1858, and the *Eamont's* crews saw many wonderful sights in that *terra incognita*.

The *Wild Dayrell* was a very similar vessel to the *Eamont*. Her illustration in the *Illustrated London News* of 1855 shows a rakish top gallant yard schooner carrying four headsails on a very long jibboom.

Her measurements were:—

Length over all	103·3 feet.	Tonnage (old measurement)	253 tons.
Beam	23·7 feet.	Tonnage (new measurement)	158 tons.
Depth	13·3 feet.		

The *Lanrick* was a beautiful little clipper brig, belonging to Jardine, Matheson, and named after Andrew Jardine's place,

Lanrick Castle. She mounted a broadside of five guns besides the midship and foc's'le head pivot Long Toms. She carried the drug between Calcutta and the Chinese ports; and raced against the *Antelope* and *Coquette* with varying results.

After this short account of the opium trade and its clippers I will now turn to China's still more important tea trade.

The Early China Tea Trade.

Until the expiration of the Company's charter in 1834, the tea trade was entirely in the hands of the East India Company. The tea was brought to England together with silk and other Chinese commodities in slow East Indiamen, which, from their lack of sailing power, were known as "tea waggons." And it was not until their charter was nearly expired that the East India Company began to build ships that could move through the water. However, in 1831-2 I find the following good times made by East Indiamen:—

The *Thames* left Canton 18th Nov., 1831.
passed Java Head 5th Dec., 1831.
arrived St. Helena 28th Jan., 1832.
arrived Portland 13th March, 1832
115 days passage.

Buckinghamshire left Canton 31st Oct., 1831, off Lizard 19th Feb., 1832.
Waterloo left Canton 31st Oct., 1831, off Lizard 19th Feb., 1832.
121 days' passage.

This is not bad work, and quite equal to that of the East India Company's successors up to the time of the American and Aberdeen clippers.

Until well into the forties the tea trade was quite small, and vessels of anything over 500 tons found a difficulty in filling their holds, but after that date, as new treaty ports were opened the tea grew more plentiful and the tea ships began to increase both in size and numbers. The first out-and-out clipper ships to visit the China Seas were Americans. They began to load tea in Canton in the early forties, and made racing passages back to New York and Boston. And they were running between China and the United States for some seven years before the first *bona fide* British clipper appeared in the East.

The American clipper, evolved in part from the Baltimore

clipper, bore no resemblance to the British clipper, each having its own particular beauty and merits.

The Yankee Clippers.

The great characteristic of the Baltimore model was the Baltimore bow. With this as his starting point, John Griffith, of New York, began, in 1841, a series of lectures on the science of ship designing, and advocated some very radical alterations in the lines and proportions of sea-going vessels. He also exhibited at the American Institute the model of a clipper built according to his theories. Though a great deal of scepticism and ridicule was showered upon his arguments he so far prevailed that, in 1843, the firm of Howland & Aspinwall commissioned Smith & Dimon, whose draughtsman Griffith had been, to build them an experimental ship on the new model.

"Rainbow."

This vessel was the *Rainbow*, of 750 tons, celebrated as the first of the great Yankee clippers. She cost, 22,500 dollars to build, or at the rate of 30 dollars a ton. Whilst she was on the stocks controversy raged round her.. Some declared, alluding to her concave bow lines, that she was turned "inside out." and was on that account contrary to the laws of Nature. Others admitted her good looks, but doubted if she could be made to sail. The chief innovation in her design was the sharpness of her ends. Instead of the full barrel-shaped bow, with the cutwater and figurehead projecting beyond the stem in what was poetically termed in old naval architecture "the sweep of the lion," she had a long, sharp, knife-like entrance with concave water lines, which carried her greatest breadth of beam very much further aft than was usual; and the heavy quarters and lumping stern were lightened and relieved by rounding up the ends of the main transom. She had an unheard of amount of dead-rise, and was far more wall-sided than any vessel of her date. As to her proportions, she was given more beams to length than was considered safe by old salts.

The *Rainbow* was launched in January, 1845, and was not long in proving a success. She did her best work on her second voyage to China, when she went out to Canton and back in 6 months

14 days, leaving New York on 1st October, 1845, she was back again in New York in April, 1846. She went out in 92 days and came home in 88 bringing to New York the news of her own arrival in the East.

Her commander, Captain John Land, who afterwards had the honour of a well-known clipper being named after him, was delighted with the *Rainbow's* qualities, and went about declaring that not only was she the fastest ship afloat, but that no ship could be built to beat her.

Rainbow is supposed to have foundered off the Horn in 1848, whilst bound from New York to Valpairaso under Captain Hayes. But though her life was a short one, she had done her work by starting a fleet of clipper ships which raised America to the level of Great Britain among maritime nations.

Captain Robert H. Waterman and "Natchez."

Whilst Captain Land was earning fame by his China passages in the *Rainbow*, another of Aspinwall's captains, Robert H. Waterman, was doing the like in a very different type of vessel, the *Natchez* one of the first New Orleans packets, built by Isaac Webb, of New York, as far back as 1831.

She was one of the first ships to be built in America with a full poop, and was a shallow-draught vessel with the old-fashioned flat bottom necessary in New Orleans packets before Captain Ends removed the Mississippi bar. This ship, which had been noted on tha Atlantic as a heavy sailing sluggard, as soon as she was acquired by Aspinwall and placed in the hands of young Bob Waterman, began a series of performances which were hailed by those who knew her as nothing short of marvellous.

On his first trip Waterman took her to the West Coast of South America, then to Canton, where he loaded tea for New York. He accomplished the whole voyage in 9 months 26 days and the passage home from China in 94 days. On his second voyage in 1844, Waterman made Valparaiso in 71 days, went up to Callao in 8 days and drove the *Natchez* across to Hong Kong in 54 days. On 15th January, 1845, he sailed from the Island of Patoe, near Macao, tea laden; rounded the Cape 39 days out; crossed the line

61 days out in longitude 31½ degrees; made the run from the equator to New York in 17½ days, and took his pilot off Barnegat on 3rd April at 6 p.m., having sailed 13,955 miles in 78 days 6 hours without having to tack once. The *Natchez* sailed her best in light winds, her best day's run being 276 miles. On his last voyage on the *Natchez* Waterman took her out direct to Hong Kong in 104 days and brought her home in 83 days.

After such voyages as these in a vessel whose previous reputation for speed had been an unusually bad one, it was only natural that young Bully Waterman should be considered one of the smartest skippers in the American Mercantile Marine, and Aspinwall determined to give his crack captain a new ship, which it was hoped would be an improved *Rainbow*.

"Sea Witch".

The result was the celebrated *Sea Witch*, a vessel which raised almost more of a stir in nautical circles than *Rainbow* had done. She was built by Smith & Dimon, of New York, the following being the details of her cost:—

For laying down, making moulds, working out, putting together frame, raising and regulating	$2,900·52
For putting in lower deck and ceiling	1,250
For putting in wales and bottom	700
For sawing	$1,500
For all other carpenter's work not included above in fitting ship for sea	$5,388·47
Total	$11,738·89

Wages at the time were two dollars per 10-hour day. Her chief measurements were:—

Registered tonnage (American)	907 $\frac{53}{95}$ tons.
,, Length	170 ft. 3 ins.
,, beam	33 ft. 11ins.
,, depth	19 ft.
Capacity for cargo	1100 tons.

Captain Waterman himself superintended every detail of her outfit, especially when it came to rigging work. He saw to it that she was fitted with skysails, royal studding sails, square lower studding sails with swinging booms, ringtails and other flying kites.

In appearance the *Sea Witch* must have been a very handsome vessel; indeed, when she sailed out of New York on her maiden voyage, she was admitted to be the most beautiful ship under the Stars and Stripes.

She had the low freeboard and raking masts of the Baltimore type, with considerable sheer, one of the outstanding features of the Yankee clipper design. Though a square-sterned ship she was built without a stern frame.

The new clipper was painted black with the old-fashioned American bright stripe, and, as she was intended for the tea trade, she was given an immense gilded Chinese dragon for a figurehead. The trailing tail of this dragon gave emphasis to her long hollow bow, which was a still further advance on that of *Rainbow*. *Sea Witch* had a sharp rise of floor (16 degrees it is said); she required a deal of ballast, indeed it is probable that she was somewhat overmasted, for she was noted for her heavy rolling and there was more than a whisper that she was unstable.

With a carefully picked crew, she left New York on the 23rd December, 1846. A strong nor'-west gale was blowing, which gave her an excellent start in life, and she ran down to the latitude of Rio in 25 days. Here she spoke the shore and sent letters back by a homeward bounder. *Sea Witch* in the hands of Bully Waterman soon proved herself to be an exceptional vessel altogether, and her records have been found very hard to beat.

Below I give the times of her first seven voyages.

First Voyage (commanded by Captain Robert H .Waterman).

From.	To.	Days Out.	Remarks.
New York	Hong Kong	104	Left New York, 23rd December, 1846.
Canton	New York	81	Off Anjer 19 days out, arrived 25th July 1847.

Second voyage (commanded by Captain Robert H. Waterman).

From.	To.	Days Out.	Remarks.
New York	Hong Kong	105	Arrived 7th November, 1847
Canton	New York	78	Left China 29th Dec.; arrived Anjer, 8th Jan. Best run 284; passed Cape 3rd Feb., 36½ days out; best run 289 in lat. 18½°, long. 80½°; crossed the line in long. 25°, 55 days out; best run 273; from lat. 5° S., long. 32° W., to lat. 21½° N., long. 60½° W., average for 8 days 275; St. Helena to Sandy Hook in 32 days; arrived New York at 9 a.m. on 15th March, 1848.

Third voyage (commanded by Captain R. H. Waterman).			
New York	Valpairaso	69	Arrived 5th July, 1848.
Valparaiso	Hong Kong	52	Arrived 7th December, 1848.
Canton	New York	79	Arrived 25th March, 1849.
Fourth voyage (commanded by Captain Frazer).			
New York	Canton	118	*Via* Valparaiso.
Canton	New York	85	73 from Java Head, arrived 7th Mar., 1850.
Fifth voyage (commanded by Captain Frazer).			
New York	San Francisco	97	Arrived 24th July, 1850. A record passage.
Canton	New York	102	Left 18th March, 1851.
Sixth voyage (commanded by Captain Frazer).			
New York	San Francisco	110	Left 1st Aug., 1851, arrived 20th Nov.
Canton	New York	110	Left 24th March, 1852.
Seventh voyage (commanded by Captain Frazer.			
New York	San Francisco	108	Arrived 8th December, 1852.
Woosung	New York	106	Left 11th March, 1853.

After her seventh voyage the *Sea Witch* gave up the San Francisco run for the direct route to China.

When homeward bound in 1855, she put into Rio with the dead body of Captain Frazer, who had been murdered at sea by his mate. After this tragedy Captain Lang took command of the, by this time, water-soaked and badly strained marvel. The end of the famous ship came on 26th March, 1856, whilst bound from Amoy to Havana with a cargo of coolies she piled up on the East Coast of Cuba and became a total loss.

And perhaps it was as well, for the coolie traffic of that date was a dreadful trade for a beautiful ship.

The Tragedy of the "Bald Eagle."

This is proved by the terrible tragedy of the clipper ship *Bald Eagle.* Like many another fine ship she gravitated into the coolie trade, and not the highest but the lowest form of coolie trade—that of carrying the refuse of China to that hell whence they never returned, the Chincha Islands. She was, in fact, but little better than a slaver. For years she drudged steadily at this awful trade, sinking lower and lower in the social scale of ships until at last a time came when even her officers were foreign, and the only sign of her past glories was the star-spangled banner which still flew from her monkey gaff.

On her last and fatal voyage her captain was a Portuguese, and he likewise shipped a crew of dagos, mostly his own countrymen the only Northerner being an Irishman, who was responsible for the terrible account of her end. How much of his yarn was an exaggeration it is impossible to say, but knowing of one or two other not dissimilar tragedies on coolie ships of that time, I should say very little.

The *Bald Eagle* was 500 miles east of Manila, bound for Callao, and reeling off an easy 10 knots under the influence of a stiff breeze. It was five bells in the afternoon watch, and all seemed quiet below, when suddenly a wild screech rang out, and the next moment an avalanche of Chinamen attempted to rush the hatchway ladders, having torn down their bunk boards for weapons. The crew, however were just in time to keep the maddened Celestials off the deck by fastening down the hatch gratings.

Then the captain, being a Portuguese acted as such, and bringing out his revolvers began shooting through the gratings at the wretched coolies, the mates following his example. But even shooting rats in a trap is sometimes dangerous, and so it proved on this occasion. The Chinese were in such a frenzy that they cared nought for the bullets, and stood out under the hatchway grating, cursing and shrieking at the shooters until there was a wriggling mass of dead and wounded Celestials piled up almost as high as the iron bars. And this was the cause of the final tragedy. So close were the pistols to the pile of dead Chinamen that a spit of flame actually set a light to the clothing of the uppermost. Immediately there was a furious rush to obtain the burning cloth, and the maddened coolies fell over one another, entirely heedless of the bullets, in their eagerness to preserve the smouldering piece. It was soon torn from the dead man's shoulder. the man who had got it at once blowing upon it to keep it alight. A bullet stopped his efforts, but another seized it only to be shot in his turn; and so the murderous business went on with the cloth still alight. As fast as those above shot down the men who held this fatal fuse others filled their places, until at last the tiny flame, which had been kept alight at such a cost, disappeared from beneath the hatch, still burning.

Half an hour later smoke began to ascend out of the fore and main hatches. The crazy Chinamen had set the ship on fire, evidently thinking that this would compel the crew to take the hatches off and thus give them their chance to rush the ship and capture her. But the Portuguese had no intention of taking any such risk. Instead they cut small holes through the deck, and began to pump water below with the aid of the wash-deck hose. With hundreds of infuriated coolies intent on keeping the flames alight, this was, of course, a useless proceeding, and in a very short while the fire had so increased that the heat and smoke compelled the Chinese to crowd under the hatchway gratings. But when they found that the crew had no intention of letting them up on deck their despair may be imagined, as it had become too late for them to be able to put out the flames themselves.

The scene now grew worthy of Dante's inferno. Beneath the bars the wretched Chinese struggled in a seething, wriggling mass of terrified humanity, packed as tight as sardines by their desperate mates further back in the heat and smoke. From this mass a long-drawn shriek of terror rang shrill and piercing into the growing darkness. To those who looked from above nought could be seen but a sea of faces turned a sickly green with fright, their eyes starting out of their heads, and their mouths opened wide as they gave vent to one horrible endless yell. As the flames approached closer and closer to the hatchways another frightful element was added to the tragedy, and that was the awful smell of burning flesh as those on the outskirts of this human maelstrom under the square of each hatch succumbed to the fire.

The crew had long since ceased to pump water, and were now only concerned in getting safely clear of the ship. The *Bald Eagle* was hove to just as night fell with great difficulty, for the smoke pouring out of the deck was so dense that the men could scarcely breathe and had to work as if in a thick fog, at the same time the deafening shrieks made it impossible to hear the orders of the officers.

By 8 o'clock the *Bald Eagle* was in a blaze fore and aft in spite of torrents of rain, which had begun to fall at sunset. Slowly the yell of the burning Chinamen had died down until a ghastly silence reigned, the last coolie having succumbed in the fiery furnace below the grim bars of the hatch gratings.

With furious haste the crew now set about launching the boats, into which they only had time to place a little biscuit and water, barely enough for one square meal. One of the boats was stove in being got over the side, so that when they at length pulled away from this awful crematorium the two quarter boats had ten men apiece, and the gunwales of the long boat were almost awash with eighteen men. The long boat had masts and sails, but the quarter boats only oars, so it was decided to tow them. The captain shaped a course for Manila. The wind was dead aft, fresh, and with a heavy following sea. All that night the long boat ran before it with the other boats in tow, all three having narrow escapes from capsizing or being swamped.

Three nights and two days were passed in this fashion, with only the nibble of a biscuit for each man and the scantiest supply of water. And, as if this was not enough, the superstitious Portuguese were terrified by the continual presence of a large shovel-nosed shark, which kept pace with the long boat, now on one side, now on the other.

On the last night the tow line of one of the quarter boats parted, and she was afterwards found stove in and floating bottom up, though there were no signs of her crew, who, it was surmised, were eaten by the shark.

Early on the morning of the third day the land was made ahead, only to be blotted out the next moment by dense mist. However, now, for the first time, the wind dropped and fell light and the two remaining boats presently found themselves entering the harbour of Manila. Here they found H.M. gunboat *Rattlesnake*, which took them on board and looked after their wants. Such was the end of one of the most horrible tales of the sea it would be possible to imagine.

American Clippers in the China Trade previous to 1850.

Though no Yankee clipper launched before the year 1850 and not many after that date, were faster than the *Sea Witch*, the following vessels were very speedy and made great reputations for themselves:—.

Date Built	Ship	Tons	Captain First voyage	Builders	Where Built	Owners
1844	*Honqua*	706	N. B. Palmer	Brown & Bell	New York	A. A. Low
1847	*Samuel Russell*	940	N. B. Palmer	Brown & Bell	New York	A. A. Low
1847	*Architect*	520	G. A. Potter	Gray	Baltimore	Nye, Parkin
1848	*Memnon*	1068	Gordon	Smith & Dimon	New York	W. Delano
1849	*Oriental*	1003	N. B. Palmer	Jacob Bell	New York	A. A. Low

The *Honqua* was named after a well-known Canton merchant who was much esteemed by Europeans on account of his integrity and kindness.

Captain N. B. Palmer left the medium clipper *Paul Jones* to take her, and he so impressed Messrs. A. A. Low & Brother, that, until his retirement, he was always given their newest ship —thus he commanded the *Honqua*, *Samuel Russell* and *Oriental* in turn. Indeed, it was said that the success of these three vessels was a great deal due to his influence on their designs and outfit. From the date that he entered the Low's employ, he seems to have acted as their confidential and expert adviser—a not unusual arrangement between owners and favourite captains.

On her maiden voyage, the *Honqua* went out to Hong Kong in 84 days and came home in 90. On the homeward passage her daily average was 158.6 miles and her best run 270. This she improved upon in 1850 under Captain McKenzie, when she broke the record from Shanghai to New York with an 88 day passage.

In January, 1848, the *Honqua* had a very nasty experience, being dismasted in a cyclone south-west of Sandalwood Island. A sudden shift of wind from S.W. to south broached her to and she soon afterwards fell into the calm centre of the storm and came very near foundering.

She met her fate eventually in a typhoon in the year 1865.

Samuel Russell was named after the founder of the firm of Russell & Co., in whose counting-house the brothers Low commenced their career. She sailed on her maiden voyage on 14th September, 1847, and went out by the Eastern Passage in 114 days, her best run being 300 miles in 38½ S., 86° E., and her log recorded:—"Strong breezes from N.W., 4 p.m. took in royal stunsails, 11.30 a.m. took in main skysail and jib-o'-jib."

In 1849 Captain Nat Palmer left her to take over the *Oriental*, and his brother T. D. Palmer took charge. The famous Captain Dumaresq went out in her and kept her log. T. D. Palmer was not a great sail carrier and Dumaresq betrayed his impatience at the sight of reefed canvas by the many exclamation marks he put after his entries in the log.

She took 89 days to Java Head, but the following will show that she was not given a fair chance.

Best day to equator 297 miles with beam wind and skysails set part of the time.

Best run after passing Cape Meridian 288 miles, "Under single and double-reefed topsails !!!!!!!!!!!!!!!!! "

1st May, "Let out reefs and made sail; consequently made a good run of 262."

6th May, "Fresh breezes, thick weather, double-reefed topsails !!!"

20th May, "Run 275 miles, fresh trades, skysails and royals in part of the time flying jib and mizen topgallantsail !!!"

But with more enterprising captains *Samuel Russell* was enabled to prove her metal. Her best run from Canton to New York was in 1852. Sailing on 6th April, she came home in 95 days, best run 328. The year before when bound home from Whampoa, she sailed 6722 miles in 30 consecutive days. She holds the record, as far as I know, between Cape San Roque and 50° south, which she did in 16 days.

In 1870, when under the command of Captain Frederick Lucas, she was wrecked in Gaspar Straits.

Of *Architect*, *Memnon* and *Oriental* I shall write later when we come to the American competition in the English trade.

The Boom in American Clipper Ship Building.

The American tea ships usually went out to China direct, but on the discovery of gold in California they were all sent round the Horn to San Franscisco, then crossed the Pacific in order to load tea home.

The Cape Horn voyage between Eastern American ports and the Californian coast the Americans have very wisely kept barred to other nations. Before the discovery of gold in the summer of 1848, it was not of much account. A few vessels went to the coast for a while before going on to China to load the season's teas, but the California trade itself was confined to the hide-droghers, who picked up their cargoes bit by bit as the supplies came down to the small coast settlements.

Dana in his *Two Years before the Mast* gives us a vivid account of this trade and the vessels employed in it. I have a record of of his ship, the *Alert*, from which it may be of interest to give a few extracts:—

"On 11th May, 1831, the *Alert* left Boston bound to Canton *via* Gibraltar. 1st June, arrived Gibraltar, 21 days out. 19th June, left Gibraltar; crossed the line 13th July, having averaged 141½ miles a day from Gibraltar. From Gibraltar to the Cape 47 days, averaging 164⅔. To Java Head 83½ days averaging 165½. To Lintin 105 days, averaging 156. Total distance by log 16,225 miles.

"Left Canton 22nd November. Passed Java Head 15½ days. Off the Cape 17th January, 56 days out. Crossed the line on 10th February. Anchored inside the Hook 7th March—105 days passage. Total log 33,579, averaging 146 per day."

The *Alert* was a 500-ton ship, with full lines and was a real specimen of the early American deep waterman at her best.

The discovery of gold in California not only woke up the Pacific coastline, but gave an extraordinary fillip to the American Mercantile Marine.

As is usual with that magic metal, the news of the find spread with amazing quickness. At first whispered rumours, and then wide-flung reports of nuggets as thick as pebbles on the seashore and gold dust to be shovelled out of the beds of mountain torrents, like so much sand, flew from continent to continent. It was the first big gold find for centuries. Men of every nationality, of every profession and of every class caught the gold fever and set out in furious haste for the great El Dorada.

There were three routes possible—round Cape Horn; by the Isthmus of Panama; and overland.

The quickest route was supposed to be by the Pacific mail steamers to Colon and across the Isthmus, but here the congestion of traffic caused endless delay, and it was often the case of the Isthmus hare and the Cape Horn tortoise, besides which this route was terribly expensive, and most gold seekers are not millionaires.

The trip overland in a prairie schooner meant facing hostile Indians and a possible death from thirst or privation. Thus it was that the greater number of the fortune-hunters chose the stormy passage round the Horn. Every sort of vessel that would float was pressed into the service, from the crack China clipper to the superannuated Indiamen, from the nimble New York pilot schooner to the war-worn veteran of the Nantucket whaling fleet. In 1849 and 1850, 760 vessels rounded the Horn from American ports alone, carrying 15,597 passengers in 1849 and 11,770 in 1850.

Many a ship entered the Golden Gate with only her pumps keeping her afloat. Many a ship, instead of anchoring, ran up on the mud flats of Mission Bay. Here, deserted by their crews, they were often taken over by complete strangers, and used as storehouses and hotels. Such a fate had the full-rig ship *Niantic,* which was floated up to what would be now the centre of San Francisco and transformed into what Westerners call a bunkhouse. A doorway was cut in her side, over which was painted:—"Rest for the weary and storage for trunks," The ship *Apollo* was turned into a saloon: whilst the hulk of the brig *Euphemia,* conveniently bilged on the opposite side of the road, was the first prison of San Francisco.

In a gold rush the prize is often to the first man on the spot, thus ship speed all at once became the desire of the whole world. Every clipper was engaged at enormous premiums. Every Down East shipyard began to work overtime. From Maine to Maryland, from Baltimore to St. John, N.B., the hammers began going night and day. Even fishing villages, where the launch of a 200-ton ship had been the sign for a general holiday and the cause of much parochial pride and rejoicing, began to build ships of 1000 tons. In some places vessels were actually built in the woods, and hauled to the water's edge by teams of oxen. Farmers turned wood sawyers, and every petty carpenter called himself a shipwright. The ships were mostly built on the share principle; the captain, the ship chandler, the block maker, sail maker and cooper each taking his proportion of shares.

And all along the Down East Coast the boys were running away to sea, until cabin boys were a glut in the market and stowing away became a necessity for an adventurous lad. The very infants learnt their knots and grew as conversant with grease and tar as the oldest shellback, whilst every Down East girl could sing "Round Cape Horn." I only know three verses; no doubt there were a hundred.

I asked a maiden by my side,
Who sighed and looked to me forlorn,
"Where is your heart?" She quick replied,
"Round Cape Horn."

I said, "I'll let your father know,"
To boys in mischief on the lawn.
They all replied, "Then you must go
"Round Cape Horn."

In fact I asked a little boy
If he could tell where he was born.
He answered with a mark of joy,
"Round Cape Horn."

New York and Boston were, of course, the two great centres of the American clipper ship boom. In New York it was said that 10,000 workmen were employed by the great shipyards along the East River. The chief firms were William H. Webb, Smith & Dimon, Jacob Bell (successor of Brown & Bell), Jacob A. Westervelt, and Roosevelt & Joyce. Of these perhaps Wm. H. Webb had the finest record. In all he built 130 vessels totalling 177,872 tons.

Boston in her turn, could boast of Sam Hall, Paul Curtis, R. E. Jackson, and Donald Mackay, one of the greatest shipbuilders that the world has known. There is no doubt that his was a peculiar genius, for certainly no mere perfection of craftsmanship could have produced his wonderful models. He never had a failure, and this is the more wonderful when we remember that a ship is something more than a building of wood and iron, that it has a life of its own, the capriciousness of which has never been better expressed than by those well-known words of Solomon when he confesses that of the four things which are too wonderful for him to understand one is "the way of a ship in the midst of the sea." Mackay's masterpieces were *Staghound*, *Flying Cloud*, *Sovereign of the Seas*, *Flying Fish*, *Westward Ho*, *Great Republic*, and the four ships built for the English Black Ball Line (James Baines & Co.), namely *Lightning*, *James Baines*, *Champion of the Seas*, and *Donald Mackay*. No vessels propelled by wind alone have ever travelled so fast through the water as the Mackay cracks.

Just before the California boom began it had been generally felt that there was considerable room for improvement in the construction of the clippers. The endeavour to give them speed at any price had made builders sacrifice other essentials, and it was soon found that the early clippers were not strongly enough put together to stand the strain of the tremendous cracking on indulged in by their daring commanders. Their bills for repairs at the end of a voyage ate a very large hole in their profits, and their cargoes were not always delivered in as good a condition as they should have been, added to which their carrying capacity compared to their tonnage was very small. So it came about that when the great demand for new clipper tonnage arrived in the wake of the gold discovery, men like Wm. H. Webb and Donald Mackay made great efforts to combine strength with speed, and in this they were imitated by the other chief builders along the coast, with the result that the American clipper ships of the early fifties were far superior to their predecessors in all-round merit.

Space will not admit a full list and description of all the beautiful clippers built in America in the early fifties, but I will attempt to give a short account of those which were best known in the China trade (*see* page 38).

American Clipper Ships Launched 1850-1851.

This list contains the names of all the foremost captains, builders, and owners in the United States; whilst amongst the ships named no less than six of them invaded the English tea trade, and caused great commotion amongst the owners of British tea ships by making passages which at the time were considered to be impossible. These were the *Celestial*, *Surprise*, *Sea Serpent*,

American Clipper Ships Launched **1850-1851.**

Ship	Tons	Length	Beam	Depth	First Captain	Builders	Owners
Celestial	860	158	34·6	—	E. C. Gardner	Wm. H. Webb	Bucklin & Crane
Surprise	1361	190	39	22	P. Dumeresq	Sam Hall	A. A. Low
Staghound	1535	215	40	21	J. Richardson	D. Mackay	Sampson & Tappan
Witchcraft	1310	—	—	—	W. C. Rogers	Paul Curtis	S. Rogers
Sea Serpent ..	1337	—	—	—	W. Howland	Geo. Raynes	Grinnell, Minturn & Co.
N. B. Palmer ..	1490	214	39	22	Low	J. A. Westervelt	A. A. Low
Flying Cloud ..	1793	225	41	21½	Creesy	D. Mackay	Grinnell, Minturn & Co.
Challenge	2006	230·6	43·6	27·6	R. H. Waterman	Wm. H. Webb	N. L. & G. Griswold
Comet	1836	229	42	22·8	Gardner	Wm. H. Webb	Bucklin & Crane
Sword Fish ..	1036	169·6	36·6	20	Babcock	Wm. H. Webb	Barclay & Livingstone
Flying Fish ..	1505	198·6	38·2	22	Nickels	D. Mackay	Sampson & Tappan
Witch of the Wave ..	1500	202	40	21	Millett	Geo. Raynes	Glidden & Williams
Nightingale ..	1066	178	36	20	——	S. Hanscom	Sampson & Tappan

Challenge, *Witch of the Wave*, and *Nightingale*. Yet if we compare them with the other seven we find that the ships that kept to the American trade were certainly faster than the ships which made such great reputations in the English trade.

The *Celestial* was Webb's first out-and-out clipper. She was a speedy little ship, though too beamy for her length. She is chiefly interesting as being one of the American clippers which entered the British tea trade. Her best performance was probably her maiden passage, when she went out to San Francisco in 104 days from New York.

The "Surprise."

From a financial, as well as a speed, point of view the *Surprise* was a most successful clipper. She was the first, also, to be fully rigged on the stocks and launched with her skysail yards across and her running gear rove off. Her launch, indeed, was made more of a ceremony than was usual in America at that date. A ladies' pavilion was specially built for it, and Hall's mould loft, gaily decorated with flags, was used as a banqueting hall, the master foreman presiding at the feast, whilst Mr. Hall provided a similar entertainment for his particular friends at his own house.

The *Surprise* was an unusually sightly vessel. If she lacked the powerful sheer and rugged appearance of strength, so marked in her Mackay rivals, she was in many respects more taking to the eye, with her graceful lines and beautifully modelled ends. These were magnificently ornamented. As a figurehead she had a golden eagle in flight, whilst the arms of New York were carved on her stern.

With regard to her measurements, 30 inches dead-rise at half floor will give some idea of her underwater body, whilst her 84-foot mainmast and 78 feet of mainyard will give an idea of her sail plan.

She carried a crew consisting of four mates, one steward, two cooks, two bos'ns, carpenter, sailmaker, four boys, six ordinary seamen, and thirty able seamen.

Captain Phillip Dumaresq, one of the most noted captains in the China trade, had her for one voyage, then he had to leave her to take over another new clipper.

When ready for sea the *Surprise* was towed round to New York by Boston's historic tug boat, the *R. B. Forbes*. There she loaded 1800 tons of cargo, valued at 200,000 dollars, for California, and it is related that her manifest was 25 feet long.

With the exception of the little *Seaman*, she was the first clipper of the season to arrive in San Francisco, having made an extraordinary run out. Though her best day's work was only 284 miles and she reefed topsails twice during the whole passage, she passed through the Golden Gate on the ninety-sixth day out, a record for that passage, which, however, was not to stand for long.

From San Francisco the *Surprise* crossed to Canton, where she was taken up to load tea for the English market at £6 a ton, double the amount offered to English ships. Her whole voyage was so successful financially that, after paying her entire receipts left her owners a clear 50,000 dollars profit.

Captain Dumaresq was succeeded by Captain Charles Ranlett, who, in turn, was succeeded by his son. Under the two Ranletts the *Surprise* put up a wonderful record racing home from China. She made six consecutive passages from Hong Kong and five from Shanghai to New York, of which the longest was only 89 days and the shortest 81, whilst she made three passages out to San Francisco averaging 109 days.

On 4th February, 1876, when commanded by Charles Ranlett, junior, she struck a sunken rock beating into Yokohama and became a total loss.

The "Staghound."

The *Staghound* was the great Donald Mackay's pioneer clipper. In her Mackay, for the first time, gave his whole thought to speed. Indeed, so little did carrying capacity enter into his calculations that the *Staghound* could barely carry her registered tonnage of deadweight.

The chief innovation in her design, which attracted the criticism of the experts, was increase of length in comparison to breadth and depth. Another point which raised the doubts of her critics was the immensity of her sail area. In all she spread upwards of 8000 yards of canvas, 1000 yards more than the usual allowance for a first-rate battleship. Her mainmast was 88 feet in length and her mainyard 86 feet.

With regard to her lines, she was the sharpest ship ever launched in Boston at that date, with a dead-rise of 40 inches at half floor, this being more than that of any of Mackay's later ships.

There was no great ceremony at the launching of *Staghound*, such as had taken place at that of *Surprise*. The 7th of December, 1850, was the date chosen, which, in accordance with her building contract, was just 60 days after the laying down of her keel. It was wintry weather, with the land frost bound and covered in snow and the harbour full of drift ice. In order to prevent the tallow freezing on the ways recourse was had to boiling whale oil. As the dog shores were knocked away, the yard foreman broke a bottle of Medford rum across her forefoot, and shouted, "*Staghound*! your name's *Staghound*!" then as she struck the water the bells of Boston pealed forth, and after waiting to see that she brought up safely to her anchors, the few frost-bitten spectators hurried home out of the cold.

In appearance the *Staghound* showed more of the points of an out-and-out racer than almost any other Mackay model. No such vessel had ever been seen in Boston before, and when she reached New York at the end of the *R. B. Forbes*' tow rope, the cautious underwriters considered that for once Mackay had overreached himself and insisted on charging extra premiums for her insurance. Nevertheless she found no difficulty in taking in a full freight at 1.40 dollars a cubic foot, which was sufficient to more than pay for her initial cost.

When she set sail for San Francisco under Captain Josiah Richardson, she carried a crew of forty-six hands before the mast, including six ordinary seamen and four boys.

She arrived at Valparaiso on 8th April, 1851, and Captain Richardson wrote the following letter to her owners:—

Gentlemen—Your ship, the *Staghound*, anchored in this port this day, after a passage of 66 days, the shortest bar one ever made here; and if we had not lost the maintopmast and all three topgallantmasts on February 6, our passage doubtless would have been the shortest ever made. The ship is yet to be built to beat the *Staghound*. Nothing that we have fallen in with yet could hold her in play. I am in love with the ship, a better sea boat or working ship or drier I never sailed.

The loss of her masts had occurred 6 days out from New York in a south-easterly gale, and notwithstanding being without a main-topsail for 9 days and topgallantsails for 12 days, she was south of the line on the twenty-first day after leaving Sandy Hook. In spite of her jury rig, she proved herself very fast in moderate breezes whilst able to log 17 miles with a fresh gale on the quarter. After repairing damages she went on to San Francisco, doing the whole passage in 107 days, her best run being 358 miles.

From San Francisco *Staghound* went on to China by way of Honolulu, making the run to Honolulu in 9 days.

Leaving Whampoa on 9th October, 1851, she made the run to New York in 94 days.

There is no doubt that the *Staghound* was either a very hard ship on her rigging, or else the gear was not sufficiently strong to stand the strain of such furious driving as her captain indulged in. This was also the case with many another famous American clipper—notably *Flying Cloud*, *Sovereign of the Seas*, *Witchcraft*, *Sea Serpent*, *Eclipse*, *Tornada* and *Comet*.

This may have been caused by a miscalculation on the part of the builders—no doubt to a certain degree it was, yet excessive carrying of sail, such as many of the American skippers delighted in, was bound to result in occasional losses of spars and sails. Anyhow *Staghound* seems to have been specially unfortunate in this respect, as on her second voyage she was obliged to put into Rio when 29 days out in order to repair damages, having been again partially dismasted. Thus, for the second time her run out was spoilt, however, she again made a good homeward passage, leaving Hong Kong on 25th September, 1852, she reached New York 95 days out.

Staghound ended her career in 1863, being burnt off the Brazil coast when bound to San Francisco from New York with a cargo of soft coal. It was said that the only thing saved from the burning ship was her United States' ensign, which her captain brought home to the owners as a relic.

The "Witchcraft."

The *Witchcraft* was Paul Curtis' first effort at an out-and-out clipper. She was a very handsome vessel, beautifully finished, with a wonderful figurehead of a Salem witch riding on a broomstick.

In point of speed, she was quite worthy to rank alongside *Surprise*, *Staghound*, and the other cracks of her year.

Her first voyage was, however, a disastrous one. Like *Staghound* she had to put into Valparaiso to replace lost spars, and on top of this she had the ill luck to run into a typhoon on her way across to China from San Francisco, when she lost her main and mizen masts. She was commanded by a very notable skipper in Captain William C. Rodgers. This man was the son of one of her owners, and, though he never served before the mast as an officer, he proved to have rare capabilities as a shipmaster.

His was a somewhat rare case. He went to sea and kept the sea for the pure love of the game, and he took command with only such experience as a few voyages to Canton and Calcutta as a passenger could give. At the outbreak of the Civil War, he was offered command in the United States Navy and served with distinction. He married a grand-daughter of the celebrated Nathaniel Bowditch.

Witchcraft was one of the select few which made the run out to San Francisco in under 100 days; this she did in 1854, taking 97 days.

The "Sea Serpent."

The *Sea Serpent* was the pioneer clipper of Messrs. Grinnell, Minturn's Californian Line and the first sharp ship built by George Raynes, of Portsmouth, N.H. She was a very rakish, thoroughbred looking racer, and the long delicate green and gold serpent forming her figurehead gave a most appropraite hint of her slippery qualities.

She was commanded by the celebrated packet captain, William Howland, the former commander of such famous Yankee packets as *Horatio*, *Ashburton*, *Henry Clay*, *Cornelius Grinnell* and *Constantine*. He was a real passenger ship captain who upheld the dignity of his position. His orders were only issued to the officer of the watch, he put on kid gloves when he came on deck and he never left the sacred planks of the poop.

On her first voyage *Sea Serpent* lost spars and sails off the Horn and was compelled to put into Valparaiso; deducting the delay so caused, her run out to San Francisco was made in 115 days. Her best passage to San Francisco was 107 days in 1853, her best passage home to New York from China 88 days.

"N. B. Palmer."

N. B. Palmer was the first American clipper to leave the ways in 1851 and the most notable of all the vessels launched from Westervelt's yard.

She was named, of course, after the celebrated Captain Palmer. A model of this beautiful ship was sent over to England and exhibited at the Crystal Palace in 1851.

In China she was known as the "Yacht," on account of the smartness with which she was kept up.

Captain Charles Porter Low, her commander, was a rich man, a younger brother of her owners, and like Captain Rodgers he went to sea because he loved the life. Captain Low and his beautiful wife made their home on the *N. B. Palmer*, and whilst in port, especially in China, they gave the most princely entertainments aboard, and many a retired British Naval Officer had pleasant memories of the crack Yankee clipper.

After a voyage or two out to San Francisco Captain Low kept the *N. B. Palmer* entirely to the China tea trade, and remained in her till she was sold and changed her flag in 1872. Her best tea passage in this time was 84 days to New York leaving Canton waters in January.

An interesting relic of the *N. B. Palmer* is still to be seen in New York. This is the carved figure of a sailor holding a compass, which stands outside the establishment of Messrs. Negus, the nautical instrument makers. This sailor at one time served as a binnacle aboard the *N. B. Palmer*. But the helmsmen used to complain that the blank stare of this wooden mariner interfered with their steering and he was eventually removed and replaced by the usual binnacle.

The "Flying Cloud."

Of all the American clippers, the *Flying Cloud* was perhaps the most notable, only one vessel, the *Andrew Jackson*, ever rivalling her double record of 89 days to San Francisco.

Donald Mackay put her on the stocks to the order of his old

friend and first patron, Enoch Train; but Train, to the regret of his life, sold her before she was launched to Grinnell, Minturn & Co.

The following are a few points worth noting in *Flying Clouds'* design and equipment. Her length of keel was 208 feet, and length over all from knighthead to taffrail 225 feet. She had 20 inches of dead-rise at half floor, which compared with the 40 inches which Mackay gave to *Staghound* his first clipper ship, shows that the great designer was already attempting to produce a fast ship with a full midship section.

Flying Cloud also seems to have had less spread to her canvas than *Staghound*, for, though their mainmasts were of equal length, the mainyard of the former only measured 82 as against *Staghound* 86 feet.

Like nearly all the Yankee clippers, *Flying Cloud* crossed three skysail yards, had royal stunsails, a reef band in her topgallant sails, four reefs, in her topsails and swinging boom and passaree to spread her fore lower stunsails and haul out the clews of the foresail.

She was placed under the command of Captain Josiah Perkins Creesy, who had the reputation of being one of the most skilful sailors of his day.

On 3rd June, 1851, the *Flying Cloud* ran out past the Hook before a light westerly air. And although the wind soon freshened to a gale, Creesy hung on to his three skysails and royal stunsails with perfect indifference to the law of the breaking strain. But there comes a time when spars begin to go, and this began as early as 6th June. The following entries in her captain's abstract log tell their own tale.

June 6.—Lost main-topsail yard and main and mizen topgallantmasts.
June 7.—Sent up topgallantmasts and yards.
June 8.—Sent up main-topsail yard and set all possible sail.
June 14.—Discovered mainmast badly sprung about a foot from the hounds and fished it.

(From this date she had doldrum weather and for four consecutive days her runs were only 101, 82, 52 and 53 miles. However she crossed the line in 21 days in spite of an unusual series of calms. Creesy, like the majority of American masters, seems to have been waging the usual belaying pin and knuckle-duster warfare with his crew, and this came to a head soon after crossing the line, some of the men being put in irons.)

July 11.—Very severe thunder and lightning. Double reefed topsails—latter part blowing a hard gale, close reefed topsails, split fore and main topmast staysails. At 1 p.m. discovered mainmast had sprung. Sent down royal and top-gallant yards and studding sail boom off lower and topsail yards to relieve the mast. Heavy sea running and shipping large quantities of water over lee rail.

July 12.—Heavy south-west gales and sea. Distance 40 miles.

July 13.—Let men out of irons in consequence of wanting their services, with the understanding that they would be taken care of on arriving at San Francisco. At 6 p.m. carried away main-topsail tye and truss band round mainmast. Single reefed topsails.

July 19.—Crossed latitude 50 south.

July 20.—At 4 a.m. close-reefed topsails and furled courses. Hard gale with thick weather and snow.

July 23.—Passed through the Straits of Le Maire. At 8 a.m. Cape Horn north 5 miles distant, the whole coast covered with snow.

July 26.—Crossed latitude 50 south in the Pacific, 7 days from same latitude in Atlantic.

July 31.—Fresh breezes and fine weather. All sail set. At 2 p.m. wind south-east. At 6 squally, in lower and topgallant studding sails. 7 p.m., in royals, 2 a.m., in foretopmast studding sail. Latter part strong gales and high sea running, ship very wet fore and aft. Distance run this day by observation 374 miles. During the squalls 18 knots of line were not sufficient to measure the rate of speed. Topgallant sails set.

August 1.—Strong gales and squally. At 6 p.m., in topgallant sails, double reefed fore and mizen topsails. Heavy sea running. At 4 a.m. made sail again. Distance 334 miles.

August 3.—Suspended first officer from duty, in consequence of his arrogating to himself the privilege of cutting up rigging contrary to my orders and long continued neglect of duty.

August 25.—Spoke barque *Amelia Pacquet* 180 days out from London bound to San Francisco.

August 29.—Lost fore-topgallant mast.

August 30.—Sent up fore-topgallant mast. Night strong and squally. 6 a.m. made South Farallones bearing north-east ½ east. 7 a.m. took a pilot. Anchored in San Francisco Harbour at 11.30 a.m. after a passage of 89 days 21 hours.

Sandy Hook to equator	21 days.
Equator to 50° south	25 „
50° South Atlantic to 50° South Pacific	7 „
50° South Pacific to equator	17 „
Equator to San Francisco	19 „
Total ..	89 days.

Flying Cloud's daily average was 222 statute miles, and her best run 374 knots in a corrected day of 24 hours 19 minutes 4 seconds. This worked into statute miles makes the 24-hour run as much as 427·5 miles. In all she sailed 17,597 statute miles at a rate of nearly 10 miles an hour.

With this triumph of *Flying Cloud's* following so closely on that of *Surprise*, Bostonians must have been jubilant, and it is likely that New Yorkers grew thoroughly tired of having the following ditty bellowed in their ears:—

> Wide-awake Down-Easters,
> No-mistake Down-Easters,
> Old Massachusetts will carry the day!

From San Francisco *Flying Cloud* went across to China. On the first day out she had a favourable whole sail breeze with smooth sea, and ran 374 miles under skysails and stunsails alow and aloft. For some reason or other a report arose in America that Creesy had died on the second day out. This stopped any action for damages which his late mate—he who "had arrogated to himself the privilege of cutting up rigging"—with the aid of a shyster lawyer, hoped to have ready for *Flying Cloud's* captain on her return. Creesy, however, was very much alive, and ran the *Flying Cloud* across to Honolulu in 12 days.

On the 6th of January, 1852, *Flying Cloud* left Canton for New York. When half way across the Indian Ocean she exchanged some Anjer fruit and vegetables for New York newpapers, in which Creesy had the pleasure of reading his own obituary notices. *Flying Cloud* arrived in New York on 10th April, after a 94-day passage. Like all Mackay's clippers she excelled in whole sail and hard breezes, and could not equal some of the smaller clippers in light winds. For instance, the *N. B. Palmer* left Canton 3 days behind *Flying Cloud*, yet arrived in New York 10 days ahead.

On her second voyage round the Horn *Flying Cloud* left New York in May, 1852. She took 30 days to the line owing to light winds. Off the coast of Brazil she fell in with her rival, the *N. B. Palmer*. There was a light northerly wind, before which the *Flying*

Cloud was running with every stitch to skysails and royal stunsails set. Whilst Creesy was ogling the sun through his sextant for noon sights, the *N. B. Palmer* was reported 6 miles ahead lying almost becalmed. The dying breeze lasted long enough to carry the *Flying Cloud* to within signalling distance of the *N. B. Palmer*, and Low, who had sailed 8 days after Creesy, reported that he had had a good run to the line, including a day's work of 396 miles. Until 4 p.m. the two ships lay side by side becalmed. As there was every appearance of a southerly breeze approaching, both ships took in their stunsails in readiness. Creesy had a fine crew this voyage, and declared afterwards that "they worked like one man and that man a hero." Low, on the other hand, had a troublesome lot, two of whom were already in irons, one for putting a bullet in his mate, and the other, one of those fighting Irishmen, for having laid out his second mate with a capstan bar.

When the wind came out of the southern horizon both ships took it at once and stood away on the starboard tack with the yards braced sharp up. From a light ripple the wind rose to a fine whole sail breeze, and as it freshened the *Flying Cloud* began to draw ahead. By daybreak next morning she had run the *N. B. Palmer* hull down to leeward. and by eight bells in the afternoon she was once more alone on the ocean. Both vessels encountered heavy westerly weather off the Horn, *Flying Cloud* eventually arriving in San Francisco in 113 days.

The *N. B. Palmer*, however, had put into Valparaiso to land her refractory seamen, after first tricing them up in the rigging and giving them four dozen apiece. This, unfortunately, cost 5 days' delay, as it gave seventeen men the opportunity to desert, and Low had some difficulty in replacing them. Thus he was 3 weeks behind *Flying Cloud* in reaching San Francisco.

This year the *Flying Cloud* took 96 days coming home from Canton, after sailing on the 1st of December.

On her third voyage she did nothing very remarkable, arriving in San Francisco 106 days out during July, 1853. But on her fourth voyage she came within two hours of her maiden passage.

Sailing from New York on the 12th January her abstract log was:—

Sandy Hook to the equator	.. ..	17 days.
Equator to 50° south	.. ..	25 „
50° South Atlantic to 50° South Pacific	.. ..	12 „
50° South Pacific to equator	.. ..	20 „
Equator to San Francisco	.. ..	15 „
Total	..	89 days.

This second record roused great enthusiasm, and Captain Creesy was feted in great style by the merchants of San Francisco, and on his return to New York he was entertained at the Astor House, then the best hotel in the city, and presented with a service of silver plate by the underwriters of New York and Boston.

On her fifth voyage in 1855, after going out to San Francisco in 108 days, *Flying Cloud* nearly ended her life in the China Seas by running on a coral reef when homeward bound with tea. However Creesy managed to float her and get her home without putting in anywhere, although she was leaking badly, having lost the shoe off her keel and had the keel itself cut through the bottom planking. For this performance he was again presented with plate by the underwriters, who reckoned that by avoiding a port of repairs he had saved them at least 30,000 dollars. It was the end of this voyage that Creesy left the *Flying Cloud* and gave up the sea.

Flying Cloud at the outbreak of the Civil War, was bought by the English Black Ball Line, and ran to Brisbane for some years. Then she descended to the Canadian lumber trade, and came to her end in September, 1873, being gutted in St. John, N.B., by fire, and her hull sold for breaking up as being not worth repairing. She was then owned in South Shields.

The "Challenge".

Shortly after the *Flying Cloud* had sailed on her maiden voyage, the notorious *Challenge* was launched at New York from Wm. H. Webb's yard on the East River. In size she surpassed any ship yet built in New York, and we are told that when lying at the foot of Pine Street her bowsprit at high tide poked over the roofs of the stores. In design she was meant to go a step further than the sharpest clipper afloat, and she had no less than 42 inches of dead rise at half floor. Her spar and sail plan was likewise tremendous, and Captain Clarke says:—"Her mainmast was 97 feet and mainyard 90 feet in length, and the lower studding sail booms were 60 feet long. With square yards and lower studding sails set the distance from boom end to boom end was 160 feet. She carried 12,780 running yards of cotton canvas, which was woven specially for her by the Colt Manufacturing Company. Her mainsail measured 80 feet on the head, 100 feet on the foot, with a drop of 47 feet 3 inches and 49 feet 6 inches on the leach."

She was painted black with a gold stripe, and, unlike other clippers, which rejoiced in white paint and varnish aloft, her masts and yards were all painted black from the trucks down.

She was one of the most costly vessels ever built of wood in America, and aroused so much interest whilst on the stocks and in the river that crowds of people visited her.

On her first passage she was commanded by the notorious Bully Waterman, who after his success in the *Sea Witch*, had arrived at the summit of his career. In New York he had been so feted and made much of that there seems little doubt that he was thrown off his balance, and became so imbued with his own importance that he was unable to put up with the lightest check to his will. The young dandy, who used to swagger down South Street in a Canton-made straw-coloured suit of raw silk and had his portrait painted by a fashionable artist, had become nothing more or less than a human tiger if we are to believe any of the thousand and one stories told of him. He had always been a desperate sail carrier, yet he seems to have become latterly a still more merciless man driver. He is said to have been the first man who padlocked his sheets and put rackings on his halliards in order to prevent his scared crew from letting things go with a run on dark nights. Yet, though this procedure gained him a name for mere reckless cracking on, he could proudly boast that in all the ships he had commanded he had never carried away a spar or called upon the underwriters for a dollar's worth of damage. Such a record no other American clipper ship commander could lay title to. The fact also that six men sailed before the mast with him through all his voyages in

Natchez and *Sea Witch* is some set off against the accusations of severe hazing and incredible cruelty which were raised against him from so many quarters. But I fear there is little doubt that these dauntless six were either privileged favourites or else too big a handful for even Bully Waterman to tackle. To this day the foc's'les of British and American sailing ships use the terrible deeds of Bully Waterman to cap the latest instance of ill usage at sea.

However much American writers may attempt to deny it the fact remains that the American Merchant Marine has always been notorious for its hard treatment of the man before the mast. Belaying pin soup had always been a institution of the Yankee packet ships, and it was continued in the clippers and other deep-watermen, though in them it was not a case of dealing with men of the toughness of packet rats, who were always ready to take advantage of the least show of softness on the part of an officer, but only the handling of a mixed crew made up of dull-witted foreigners and landsmen. For when that rarity, a real British or American seaman was found in a clipper, he was soon settled one way or the other—he either stood up to the mates and cowed them into leaving him alone, or he lost the number of his mess by what was called an accident in the log book. And if this was the fate of the man who knew his work and had spirit in him it may well be imagined what was the lot of the wretched Dagoes and Dutchmen and the still more unfortunate Shanghai-ed landsmen. Without mincing words it was just sheer, undiluted hell with the lid off. It was not only work until you drop, but get up and go on working after you have dropped under the blow of an iron belaying pin.

The American ships were noted for their smart appearance.

> A Yankee ship comes down the river,
> Blow, boys, blow!
> Her masts and yards they shine like silver,
> Blow, my bully boys, blow.

goes the well-known shanty.

No wonder masts and yards shone like silver when it was the custom on some of these clippers to have both watches scrubbing yards with sand and canvas on moonlight nights. Then the pride that a Down-East mate took in his deck made a day watch below almost an unknown luxury on many a ship flying the Stars and Stripes. Down at their prayers in the slush and pulp would be both watches hour after hour mechanically working the stones, and not a man dared straighten his back for a second unless he chose to risk being crippled for life by a heavy handspike. The usual Yankee mate's method of dressing a deck was first to scrape and holystone very thoroughly, then give a coat of coal tar, then scrape again and finally holystone until the deck came up as white as snow. And how well the old hand knew those ships with shining yards and decks of snow! How he avoided such ships as if they were plague-ridden. He well knew that such smart results were not obtained by fair methods but by the sheer brutal driving of worn out men.

I have been shipmates with men who had sailed for years in Yankee clippers and Cape Horners, and they were almost invariably poor sailormen yet incomparable scrubbers and swabbers, and whether they were big men or little, it mattered not, there was no spirit left in them—it had been broken long ago.

But to return to Bully Waterman, when he took command of the *Challenge*, his reputation was so bad that no real sailorman would sign with him of his own free will. In the *Sea Witch* he had made a name as a pistol shot, and his fondness for potting men on the yards had compelled him to make a practice of leaving his ship before she anchored.

One of the stories told of Waterman related to Fraser who, after being mate, succeeded him in command of the *Sea Witch*. Frazer is said to have bluffed him in his own cabin. The story goes that whilst they were below together, Frazer produced two bull-dog pistols and pushing them across the table to Bully Waterman, said; "Either you or I have got to leave this ship."

At which Waterman at once knuckled under with the remark: "You are the only man I ever had any respect for."

It was always said of Waterman that at the start of a voyage, as soon as the ship was clear of the shore-goers, he would call to his steward: "Bring me a bucket of salt water to wash off my shore face," and straightway he would change from a priggish sanctimonious soft-mouthed humbug to a regular fiend in human shape.

It would be impossible to detail all the horrors he is supposed to have committed, from casting off the lee main-brace in a Cape Horn snorter and jerking half-a-dozen men into the sea to shooting his own child.

The *Challenge* sailed from Pier 19, foot of Wall Street in July, 1851, just a month later than *Flying Cloud.* Her proud owners went out with her as far as the Hook, and before they left the ship Bully Waterman mustered his crew aft. And whilst the "old man" was making the usual speech, lauding the merits of his ship for good grub and little work, with the usual sting in its tail as to the hell he would make it under certain contingencies, his officers were ransacking the foc's'le for black bottles, bowie knives, and bull-dog pistols. At the same time the carpenter was employed breaking the points off the men's sheath knives.

It was soon found that out of a crew of fifty-six men and eight boys there was scarcely a real sailorman, owing to the captain's reputation. Only six men could steer and only four could speak English. With all their tricks the crimps had only been able to collect a rabble of Dagos, Dutchmen, Southwegians and niggers, of whom many were sick and the majority dead-beats.

At the sight of these sweepings from all nations, the owners seriously proposed that the *Challenge* should put back for a fresh crew, but Waterman would have none of it. They were just the sort of foc's'le crowd which gave him a chance to show his talents.

"I'll make sailors of 'em or else mincemeat," he growled. And it was not long before he began to make good his words The owners, N. L. and G. Griswold (known in New York as No Loss and Great Gain Griswold) had hardly left the ship before he laid open the darkey steward's scalp with a carving knife. From that moment never a day passed without blood on the decks

The mate, Jim Douglas, a big block of a man, was as tough a bucko as the old man and as headstrong.

Another nasty man to tackle was the Swedish carpenter. He actually had the hardihood to fall foul of his captain before the *Challenge* was out of sight of Sandy Hook.

"You're my boss here," grunted Ships, "but, by gar, if I had you ashore, I'd lick you."

"By the powers, we'll try it," responded the "old man," and lugged off his coat.

It was a fair fight and the carpenter had the best of it. From that hour Bully Waterman left him alone.

In spite of the number of her crew, the *Challenge* soon grew shorthanded, and her sailroom had to be cleared out to house the sick men, of whom five died and eight were still in their bunks on the ship's arrival at San Francisco.

There were five passengers on board, and they must have had a lively passage. Off Rio one day they came on deck at eight bells in order to witness a muster of all hands and their dunnage by the mate.

Kicking open the chests, Jim Douglas hove their contents out on the deck. In those of the four English-speaking seamen, he discovered, amongst the usual gear, four "prickers". Pouncing upon these, he held them up to view and sneered: "So you've been stealing from other ships! I'll take care of these."

One of the four was an old man-of-war's man, and this was more than he could stand. With a grunt of rage he went for the bucko and knocked him flat.

Whilst this was going on Bully Waterman was up on the poop trying to catch the sun with an instrument called a "circle of reflection," As soon as he saw his mate floored, he jumped off the poop into the midst of the crew, and struck out right and left with his circle of reflection until he had hopelessly smashed the instrument. Another account says that he killed two men with an iron belaying pin, and that Douglas received no less than twelve knife wounds. The man-of-war's man however, disappeared in the general mix-up.

"He's gone over the rail," declared the men.

"D—n him, I knew he was afraid of me," growled the mate. The other three English-speaking foremast hands were eventually seized, taken into the cabin and made to sign a statement in the presence of the passengers. After which they were tied up by their thumbs to the mizen rigging and flogged. But the man-of-war's man was not found until the ship was off the Horn. The mate, who did not believe the man had gone overboard, waited for an opportunity to search the foc's'le whilst all hands were on deck. He chose a night when the *Challenge* was hove to under Staten Island during a heavy gale. All hands were called aft and made to stand by on the poop. This was Douglas's opportunity. Taking one of the ship's boys with him, he went forward and carefully searched the foc's'le with a lantern.

Between the lower bunks and the floor, boards had been nailed up enclosing the space underneath. These boards he carefully felt over with his hand. At last he found what he was looking for—a couple of knots, showing that there was a becket on the inside, and as soon as he cut the knots the plank gave way and fell out. The boy was then ordered to crawl into the space under the bunks and feel round for the missing man.. It was not a pleasant job for the boy, but he was more afraid of the mate than the devil himself, so in he went.

Presently his hand touched something warm! In a moment he was out of the hole and scrambling up the foc's'le ladder, screeching with fright.

And Mr. Bucko Douglas caught the panic, and, dropping his lantern, followed the boy; but as soon as he reached the deck, he pulled himself together and waited.

In a few seconds the missing man appeared, but on seeing Douglas dropped on his knees and begged for mercy.

Snarling out a curse, the mate struck the miserable wretch with a heaver. The latter tried to protect his head, with the result that his arm was broken in two places.

He was then put in irons and kept a prisoner on bread and water for the remainder of the passage, his broken arm mending as best as it could.

Bully Waterman, like his mate, had a partiality for a heaver and generally carried one tied to his wrist, just as a New York policeman carries his club. The usual victims of the "old man's" heaver were the helmsmen; he used to make a practice of standing behind them until he saw an opportunity of using it. One night he beat three men into unconsciousness, one after the other—the first for having dirty hands, and the other two for not understanding the compass.

The second mate of the *Challenge*, a man named Cole, was just such another hard nut as Bully Waterman and Jim Douglas.

Whilst off the Horn he was up aloft with the men trying to furl the mizen topsail, which was giving trouble and blowing up over the yard; this made it dangerous to move out along the yardarm.

"Get those men out," yelled the impatient mate from below; "lay out on the yard there."

The men were almost frozen and utterly done up, so that the thrashing sail was altogether too much for them. The mate's words, however, so infuriated Cole that he lost all control of himself. Springing on to the yard and holding on by the tye, he booted three men off the weather foot-ropes.

Two of them hit the brace bumpkin and rebounded into the sea, where they floated for a short time without any attempt being made to save them, and then sank. The third fell on the poop and lay there groaning. The angry fury of a mate immediately pounced upon him.

"Hi, you!" he cried, "why ain't you dead? You are dead." Then turning to some of the men on deck, he asked: "Say? Has this fellow got a blanket? Yes! then bring it up."

And there and then the mate stitched the living man up in his own blanket and had him tossed overboard—*still groaning.*

The *Challenge* took 55 days from Sandy Hook to the Horn. The crew by this time had been so weakened by sickness and knock-outs that the yards threatened to take charge every time the braces were started in dirty weather. It was perhaps lucky for the clipper that with the exception of a bad spell of westerlies rounding the corner, she had a passage of moderate winds and fine weather.

She was barely on the other side of the Horn before the Fates got even with the second mate.

Cole was bracing the yards with the cold eye of his skipper watching him from the poop. Somehow he managed to displease the "old man," who, after warming the air with his language, suddenly leapt for his second mate, springing upon him from the poop in his usual headlong style. But Cole was the quicker with his fists and landed his captain a blow between the eyes which sent him to the deck. The mate then took a hand and hurled a belaying pin at his formidable junior; but though as a rule no man is a finer shot with his missile than an American mate, this time Douglas missed and instead of the greenheart pin downing the second mate it went clean through the topgallant rail. Cole was evidently a nasty man to tackle, for the redoubtable Jimmy Douglas, on missing his shot, thought it wise to take to his heels. The second mate, however, soon caught him and knocked him flying into the fore-topsail halliards.

Having now vanquished his two superior officers the valiant Cole faced the crew and shouted:

"Here's the ship! take her if you want her."

But nobody moved; that was where the safety of a mixed crowd came in! Cole sized up the state of things at a glance, and with admirable *sangfroid* quietly resumed his place as officer of the watch as if nothing had happened.

About an hour later the old man, having recovered from his knock-out, came on deck again.

Going up to his second mate, he said:

"Come into the cabin, Mister Cole, I want to see you about something."

Under the circumstances it was not unnatural that Cole should smell a rat.

"I'm officer of the deck and won't leave it for any man," he replied.

Then Bully Waterman began to wheedle.

"I'll give you my word nothing will happen, Mr. Cole, I want to talk to you. I've forgotten that other trouble."

Cole was at length fooled into following his captain below. He came up again in about an hour and it was thought that the row had been patched up.

But although he stood his watch out, and again came on duty for the six to eight dog-watch, before the night was out he was too weak too stand up without help.

The "old man" had given him a drugged glass of grog. Cole saw his own finish, but getting hold of two of the ship's boys, he made them support him whilst he dragged himself forward, with the idea of hiding from Bully Waterman. But the latter was on the watch, and as the two boys were struggling to get the wretched second mate forward, he slipped up behind him and struck him on the head with his ever-ready heaver. Cole was then ironed, hands and feet, and thrown into the port quarter boat, where he was kept for the rest of the passage and fed on bread and water like the unfortunate man-of-war's man.

But it was the last outrage, the murder of a poor old Italian, which caused the San Francisco trial.

"Old Papa," as he was called by the men, one day failed to show up on watch. It was the afternoon watch, and Bully Waterman, like a tiger in search of blood, accompanied his mate forward in order to fetch the Dago out of the foc's'le. Pounding on the scuttle with his heaver, the irate skipper yelled for Old Papa, and when the man appeared asked him why he was not standing watch. By way of reply Old Papa mumbled in Italian and pointed to his feet. They were black with mortification! Someone had stolen his only pair of boots, and he had been compelled to go barefoot through the bitter weather of the Horn, with the result that his feet had been frozen.

But as soon as he began to mumble Italian, Bully Waterman let fly.

"Curse you, speak English, can't you?" he yelled and straightway struck the wretched old man over the head with his heaver.

The Dago dropped as if he had been pole-axed.

At this, the captain roared to the steward to bring some hot whisky forward.

"You don't need any whisky," said the mate calmly, "the man's dead."

The *Challenge* was 34 days running up to the line and another 19 on to San Francisco. Her best day's run during the passage was 336 miles under all plain sail with wind abeam. On the whole her performance was disappointing, yet Bully Waterman had spared no effort in order to send her along. It is said that he never took his clothes off except to change them, and made a practice of snatching what sleep he could on a settee in his chart-room.

As was customary at that time with hell-ship captains, Waterman slipped ashore before the *Challenge* came to an anchor. And it was well for himself that he did so, for, as soon as the atrocities of the passage became known in San Francisco, a mob of red-shirted miners collected in order to lynch him and his officers.

They first marched down to the Pacific Wharf where the *Challenge* was moored, but found no one aboard except old Captain John Land of *Rainbow* fame, who had been placed in charge by the ship's agents. Putting him in their midst the crowd, now some two thousand strong, marched on Alsop & Co., the agents—but just as they were forcing the locked doors with crowbars, the Monumental Fire Engine House began to ring its bell in order to call out the Vigilance Committee. Then the Marshall appeared and with some difficulty managed to cool the wild temper of the mob, who presently dispersed after having received a promise that justice should be done.

Bully Waterman, as soon as the first excitement was over, gave himself up. He was placed under a 50,000 dollar bond, which was soon raised to 100,000 dollars. Jim Douglas was also arrested. Both men had no difficulty in squaring judges and jury at the trial, and so escaped punishment. The case, however, was too notorious for Bully Waterman ever to go to sea again. But, fortunately for himself, he was very well off, and already possessed a holding in Solano County, California, where he is said to have founded the city of Fairfield.

For many years he held the office of Port Warden and Inspector of Halls in San Francisco. Like many another man-driver at sea, he was another being ashore, and took up religion with so much zeal that he actually went about "saving souls." On one occasion, when employed on this voluntary missionary work, he boarded a ship lying in San Francisco Bay, but, unfortunately for the old bucko, some men who had sailed with him happened to be among the crew. These men threw him overboard, and were busy trying to drown him by shoving him under the water with a long pole when the harbour police rescued him. He died on his farm in 1884 at the age of seventy-six.

This first passage of the *Challenge* may seem too highly coloured —many may consider my account a great exaggeration—but I fear it is only too true, and those who have any acquaintance with the American Cape Horn fleet as it used to be, will not find anything very unusual in the doings of Bully Waterman and his bucko officers.

Many of the incidents I have related were sworn to at the trial, yet they could not convict in the face of Waterman's well laid out dollars. But it is a significant fact that neither the owners of the *Challenge* not the underwriters ever said a word in defence of Waterman.

With a new captain, officers, and crew the *Challenge* made a fine run across to Shanghai and loaded tea for England. There she was so much admired that her lines were taken off for the Admiralty whilst she lay in the Blackwall Dock. Her subsequent career is shrouded in mystery, owing to later clippers being given the same name.

The "Comet."

The *Comet* was another of the larger American clippers. She made some very fine passages under Captain Gardner, the best being—

1852.—New York to San Francisco 102 days.
Canton to New York 99 days.
1853.—New York to San Francisco 112 days.
(After losing fore topmast and main topgallantmast in a cyclone off Bermuda.)
San Francisco to New York 76 days.
1854.—Liverpool to Hong Kong 84 days.
(Averaging 212 miles a day.)

She was afterwards sold to James Baines, of the Black Ball Line to Australia, and renamed *Fiery Star*.

On 1st April, 1855, she left Moreton Bay for London. On the 19th one of the men reported a strong smell of smoke in the foc's'le, which soon burst forth in clouds, the fire being located in the lower hold. The captain (Yule) immediately had all hatchways battened down and ventilation pipes blocked up. The ship was running free 400 miles from Chatham Island. A few days before a heavy sea had made matchwood of two of the boats, so the weather was evidently true casting weather.

On the 20th a steam pump was rigged down a hatchway, and wetted sails were fastened down over all vents in the deck. But the fire continued to gain, and at 6 p.m. it burst through the port bow and waterways. The four remaining boats were at once provisioned and got over the side. Seeing that there was not room for everybody, Mr. Sargeant, the chief officer, four A.B.'s, and thirteen boys agreed to stand by the ship, the remainder of the passengers and crew to the number of seventy-eight leaving in the boats under the captain.

As soon as the boats had left, Mr. Sargeant renewed every effort to subdue the fire, and at the same time altered his course in order to get into the track of other ships. Then for 21 days he and his gallant band fought the flames and the numerous gales of those regions. Finally, on 11th May, when the foremast was almost burnt through, a ship. called the *Dauntless*, hove in sight and took the men off the doomed ship.

The people of Auckland, New Zealand, whither the *Dauntless* was bound, for their gallantry in remaining behind presented Sargeant and his brave crew with a testimonial in the shape of £160, £80 going to the mate and £80 to the crew.

The captain, with the four heavily-laden boats, was also rescued after experiencing all the usual hardships of hunger, thirst, and bitter weather.

The "Swordfish."

The *Swordfish* was the third of the famous clippers built by Wm. H. Webb in 1851. Though only half the tonnage and not nearly so sharp-ended as the *Challenge* and *Comet*, she was generally acknowledged to be Webb's masterpiece.

She was commanded on her first voyage by Captain David S. Babcock, the brother-in-law of the famous Captain N. B. Palmer, and a member of a very distinguished New England family.

On her first passage out to San Francisco it was arranged that the *Swordfish* should race the new Mackay crack *Flying Fish* for large stakes. The abstracts of this race are worth recording.

Swordfish	Out.
Nov. 11, 1851.—Left New York	
Dec. 4, 1851.—Crossed equator	23 days.
Dec. 26, 1851.—Crossed parallel 50° S.	45 „
Jan. 3, 1852.—Crossed parallel 50° S. (Pacific)	53 „
Jan. 22, 1852.—Crossed equator (Pacific) ..	72 „
Feb. 10, 1852.—Arrived San Francisco	92 „

Flying Fish.	Out.
Nov. 11, 1851.—Left Boston.	
Nov. 30, 1851.—Crossed equator	19 days.
Dec. 26, 1851.—Crossed parallel 50° S.	45 „
Jan. 2, 1852.—Crossed parallel 50° S. (Pacific)	52 „
Jan. 26, 1852.—Crossed equator (Pacific) ..	76 „
Feb. 17, 1852.—Arrived San Francisco	98 „

It will be seen that both vessels made the run in under 100 days, *Swordfish's* time being within three days of the record of *Flying Cloud*, and the best of that year.

From San Francisco *Swordfish* crossed to China, and, loading tea at Canton, left on 26th September, and made the run home to New York in 90 days.

On her second voyage *Swordfish* was taken over by Captain Charles Collins, and arrived in San Francisco on 30th May, 1853, 105 days out.
From San Francisco she went across to Shanghai in 32 days 9 hours. Below I give the abstract of her log, which it will be of interest to compare with those of the later British tea clippers.

Log of Clipper Ship "Swordfish" from San Francisco to Shanghai.

June 17, 1853.—Lat. 35° 25′ N., long. 126° 35′ W.; pilot boat left 2 p.m.; 4 p.m., lost use of main topgallantsail, stay parted; foggy. Dist. 236 miles.

June 18.—Lat. 32° 30′ N., long. 132° 7′ W. First part clear, ends foggy. Dist. 340 miles.

June 19.—Lat. 30° 36′ N., long. 137° 34′ W. Fair breeze, hazy. Dist. 280 miles.

June 20.—Lat. 28° 40′N., long. 140° 49′ W. Fair breeze, hazy. Dist. 250 miles.

June 21.—Lat. 26° 53′ N., long. 144° 23′ W. Pleasant trades. Dist. 225 miles.

June 22.—Lat. 25° 25′ N., long. 147° 46′ W., Pleasant trades. Dist. 202 miles.

June 23.—Lat. 23° 56′ N., long. 151° 14′ W. Pleasant trades. Dist. 201 miles.

June 24.—Lat. 22° 49′ N., long. 153° 27′ W. Light airs and calms. Dist. 142 miles.

June 25.—Lat. 21° 30′ N., long. 156° 40′ W. Light airs; 5 a.m., "land, ho!" Morree Island. Dist. 208. Total distance run 2084 miles—average per day 232 miles.

June 26.—Lat. 20° 5′ N., long. 160° 15′ W. Light breeze. 2 p.m., in the passage of the islands; passage 8 days 2 hours.

June 27.—Lat. 18° 33′ N., long. 162° 46′ W. Very light airs. Dist. 180 miles.

June 28.—Lat. 18° 34′ N., long. 166° W. Very light airs. Dist. 181 miles.

June 29.—Lat. 18° 37′ N., long. 170° 4′ W. Good breeze. Dist. 240 miles.

June 30.—Lat. 18° 37′ N., long. 173° 21′ W. Calm and light air. Dist. 190 miles.

July 1.—Lat. 18° 50′ N., long. 176° 48′ W. Bent old sails; ship does not sail as fast as with heavy suit; 1½ knot difference by log.

July 2.—Lat. 18° 38′ N., long. 180° W. Light trades on meridian. Dist. 195 miles.

July 4.—Lat. 18° 38′ N., long 176° E. Fine trades. Dist. 230 miles.

July 5.—Lat. 18° 43′ N., long. 172° 51′ E. Light trades. Dist. 190 miles.

July 6.—Lat. 18° 47′ N., long. 169° 16′ E. Light trades. Squally. Dist. 212 miles.

July 7.—Lat. 18° 52′ N., long. 165° 29′ E. Fair trades. Heavy swell. Dist. 228 miles.

July 8.—Lat 18° 49′ N., long. 161° 53′ E. Fair trades. Heavy swell. Dist. 210 miles.

July 9.—Lat. 18° 42′ N., long. 157° 25′ E. Fair trades. Pleasant. Dist. 226 miles.

July 10.—Lat. 18° 35′ N., long. 154° 38′ E. Light airs. Dist. 157 miles.

July 11.—Lat. 18° 25′ N., long. 150° 27′ E. Light airs. Hot and sultry. Dist. 222 miles.

July 12.—Lat.. 18° 19′ N., long. 146° 54′ E. Light airs. Hot and sultry. Ends squally. Dist. 229 miles.

July 13.—Lat. 18° 20′ N., long. 143° 28′ E. Light airs. 5 p.m., "land, ho!" Islands of Pagon and Almaguan (Ladrones), 8 p.m. passed through all clear. Dist. 210 miles.

July 14—Lat. 18° 19′ N., long. 139° 57′ E. Begins light air, ends squally. Dist. 210 miles.

July 15.—Lat. 19° 27′ N., long. 135° 38′ E. Squally, much rain, thunder and lightning. Dist. 265 miles.

July 16.—Lat. 21° 4′ N., long. 131° E. Commences very warm. 2 a.m., sharp chain lightning; looks very bad; expect a typhoon; in all sail except close-reefed fore and main topsails; battened down all hatches. Daylight, strong breeze; overhead clear; horizon foul and looks bad; this maybe caused by the ship drawing in between N.E. trades and S.W. monsoons. The ship went 9 knots, wind abeam under two close-reefed topsails, made sail as required. Dist. 260 miles.

July 18.—Lat. 27° 28′ N., long. 125° 14′ E., Strong breezes. Midnight all sail. Dist. 253 miles.

July 19.—Lat. 30° 50′ N., long. no observation. 36 hours in this day. 11.30 a.m., made Saddle Island. 11 p.m., anchored for daylight off Gutztaff Island (Shanghai entrance). Dist. 224 miles. Daylight took Shanghai pilot and proceeded up Yang-tse-kiang.

The total distance run was 7100 miles, giving an average per day of 225 miles, and a passage of 32 days 9 hours. (It will be noticed that 3rd July and 17th July are omitted in the captain's abstract).

On the following voyage, on the usual round of California, China and home, the *Swordfish* under Captain H. N. Osgood logged 39,977 miles, averaging 153 miles a day, and made the round in 10 months and 10 days, including 35 days in port.

"Flying Fish."

The *Flying Fish*, *Swordfish's* rival in 1852, was one of the fastest and most beautiful clippers designed by Donald Mackay, and her records were very nearly as good as those of the celebrated *Flying Cloud.*

In 1853 she won a magnificent race out to San Francisco against the *John Gilpin.* The *John Gilpin* was designed by Boston's other great shipbuilder, Sam Hall, and measured 1089 tons, thus like the *Swordfish* she was handicapped by tonnage. Under Captain Doane, she sailed from New York on 29th October, 1852, and was followed by *Flying Fish* on 1st November. It was the best season of the year for making good passages, and both vessels made splendid runs south. The *Flying Fish* was actually down to the parallel of 5° N. on the sixteenth day out from New York, and the following day lost her wind in 4° S., 34° W. Here Nickels made a great mistake for, instead of standing boldly on and trusting to slants to carry him clear of San Roque, he did just what the great Maury advised captains not to do: he tried to work to the eastward against the westerly set and was thus held up in 3° N. by calms and doldrums for four whole days. However, this was his only real halt in the whole passage. On 24th November in 5° S., the *Flying Fish* and *John Gilpin* were level with each other, the latter some 37 miles more to the eastward, and the *Flying Fish* so close on the land that she

[illegible]y slant to stand off shore. The *John*
[illegible]r working S. gained three days on
[illegible] Roque and 50° south, but this the
[illegible]ant through the Straits of Le Maire.
[illegible]sels were in company for the first
[illegible]ls actually invited Doane to come
[illegible]he Horn, however, does not allow
[illegible]rked in his log: "I was reluctantly
[illegible]n."

[illegible] westerlies between 50° S. and 50°
[illegible] ship had a great advantage, and
[illegible]d four days on her rival.

[illegible]oth ships crossed the parallel of
[illegible]company with another very fast
c[illegible]d sailed from New York on 12th
O[illegible]nate with her winds.

[illegible], the *Flying Fish* crossed the
equ[illegible]ad of the *Wild Pigeon*, and 260
mile[illegible]h was also further to the westward[illegible] the *Westward Ho*, another Mackay clipper
owned by Sampson & Tappan, and of almost the same tonnage as
Flying Fish, crossed the line 4° further west, and the two ships of
almost the same tonnage and design (the *Westward Ho* being on her
maiden passage) and owned by the same men, made a dead-heat race
of it to San Francisco, arriving on 1st February. *Westward Ho*, however, had taken 103 days on the passage, having sailed from Boston
on 20th October. She had been badly handicapped by a drunken
captain, as the following account given by one of her passengers,
a seaman, will show:—

"*Westward Ho* ought to have done the run in 90 days. The
captain was a drunken beast and remained in his cabin for nearly
the whole passage, boosing on his own liquor and that of the
passengers from whom he could beg, and at last broke out the forehold
in search of liquor, and found some champagne cider on which he
boosed the remainder of the passage. We were off the River Plate
with a fair strong wind, headed east and north for several days,
until there was nearly a mutiny among the passengers. I finally

told the mate to put her on her course and we would back him up in any trouble. The captain never knew of any change; we lost at least 10 days by such delays. At one time after passing Cape Horn we were running about N. by W., wind S.S.W., long easy sea and wind strong under topgallant sails, and she was going like a scared dog, her starboard plank sheer even with the water, two men at the wheel and they had all they could do to hold her on her course. One day she ran over 400 knots—17 knots per hour—another day she ran 388 knots. The drunken captain was at once displaced in Frisco, and the mate, who had navigated from Boston, placed in charge. He made the run to Manila in 31 days.'

The *John Gilpin* crossed the line two days behind *Flying Fish* and *Westward Ho*, making a wonderful run in, just failed to catch her rival.

The abstracts of this fine race were:—

Flying Fish.		Out.
Nov. 1, 1852.	—Left New York.	
Nov. 22, 1852.	—Crossed equator	21 days.
Nov. 24, 1852.	—In 5° S.	
Dec. 19, 1852.	—Crossed 50° S. (Atlantic) ..	48 „
Dec. 26, 1852.	—Crossed 50° S. (Pacific)	55 „
Dec. 30, 1852.	—In 35° S.	
Jan. 13, 1853.	—Crossed equator	73 „
Feb. 1, 1853.	—Arrived San Francisco	92 „

John Gilpin.		Out.
Oct. 29, 1852.	—Left New York.	
Nov. 22, 1852.	—Crossed equator	24 days.
Nov. 24, 1852.	—In 5° S.	
Dec. 15, 1852.	—Crossed 50° S. (Atlantic)	47 „
Dec. 26, 1852.	—Crossed 50° S. (Pacific)	58 „
Dec. 30, 1852.	—In 35° S.	
Jan. 15, 1853.	—Crossed equator	78 „
Feb. 1, 1853.	—Arrived San Francisco	94 „

The *Flying Fish*, like most of Mackay's powerful designs, did better in the Californian trade than in the China trade, and the following list of her passages to San Francisco bears comparison with the work of any other American crack.

1852— 98 days.	1855—105 days.
1853— 92 „	1856—113 „
1854—113 „	1857—100 „
1855—109 „	

In November, 1858, the *Flying Fish* was wrecked on her way out of the Min River, loaded with Foochow tea for New York. She was abandoned by the underwriters, and the wreck sold to a Spanish merchant of Manila, who managed to float her, and then had her practically rebuilt at Whampoa. Then for some years she sailed between Manila and Cadiz, disguised under the name of *El Bueno Suceso*, and eventually foundered in the China Seas.

The "Witch of the Wave" and "Nightingale."

These are chiefly celebrated for their performances in the English tea trade. They were of the smaller class of American clipper and, unlike the *Flying Fish*, were more suited to the China Seas than the Southern Ocean, thus neither of them did anything remarkable in the Californian trade.

Witch of the Wave was the pride of the interesting old port of Salem, where she was owned. She was an extreme clipper, with 40 inches of dead rise, 81 feet of mainyard, and a 90-foot mainmast.

The *Nightingale* had a somewhat curious history. She was originally built to carry passengers to the World's Fair in London, and then to be exhibited in the Thames, as a typical American clipper ship. She was, therefore, fitted out regardless of expense with large saloons and the most luxurious cabin arrangements. And in every other way she was a most expensively built vessel, her figurehead, for instance, being a most beautifully carved bust of the famous singer, Jenny Lind, in honour of whom the vessel was named. Unfortunately for the promoters of the scheme, they fell short of money before the ship was launched, and the *Nightingale* was then bought by Sampson & Tappan for 75,000 dollars.

After she had had a very successful career in the British tea trade, of which we shall hear later, she was sold to the Brazilians, who put her into the African slave trade. After two or three years in this horrible traffic she was captured, about 1860, by an American gun boat, and sent home as a prize. Then, when the Civil War broke out, she was fitted as an armed cruiser. At the close of the war she went back to the Californian and China trade, and years later, ended her days under the Norwegian flag.

American Tea Passages 1851-1853.

On the following page will be found a table showing the passages from China to America at the zenith of the American tea trade. Also the performances of other clippers, of which lack of space has not permitted a description.

The Rivalry of Great Britain and America in the Tea Trade.

At the present day, when the supremacy of our Mercantile Marine as the world's carrier is so firmly established, it is hard to realise that in the forties and fifties America was not only our equal, but in many ways our superior. In mere numbers the two great maritime nations were about equal. Our hard wood ships lasted longer than their soft wood ships, but it must be remembered that the former could not be built for less than £15 per ton, whereas the latter could be built for £12 per ton and even less. In model and design we had no ships that could compare with such vessels as the *Sea Witch*, *Honqua*, *Samuel Russell*, and *Oriental*. Again, in the cut and set of their sails the Americans were first and the British, along with the rest of the world, nowhere. And American ships worked several men lighter than British ships of equal tonnage owing to their use of deck winches, patent sheaves, light Manila running gear, and large blocks, where we were content with common sheaves, stiff hemp gear, and the hard worked handy billy.

But there were very good reasons for this lack of enterprise on our part. The chief of these reasons were our prehistoric tonnage and navigation laws.

The tonnage law—which dated from 1773, and was not radically altered until 1854—encouraged a very bad, slow type of ship, as only length and breadth were taxed and not depth, thus the usual cargo carrier was far too short and too deep, and ships of proper dimensions were so handicapped by tonnage dues that they were practically non-existent except in the rich opium trade. As soon as this law was revoked, and Moorson's plan of internal measurement—the outcome of which is our present registered tonnage—was adopted, it was at once found that the alteration was more advantageous to the shipowner than the old law, besides giving the builder more latitude to show his talents.

American Tea Trade Passages 1851-1853.

Ship	Ton'ge	Captain	From	Sailing Date	To	Days
				1851		
Celestial ..	860	E. C. Gardner	Woosung	Mar. 4	New York	117
Mandarin ..	776	Stoddard	Canton	April 4	,,	118
Sea Witch ..	890	Frazer	,,	Mar. 18	,,	102
Honqua	706	——	Woosung	Aug. 19	,,	129
Staghound ..	1535	J. Richardson	Canton	Oct. 9	,,	94
Sea Serpent ..	1337	W. Howland	,,	Oct. 14	,,	101
Panama ..	670	——	Woosung	Oct. 17	,,	108
Gazelle	1244	Henderson	Canton	Dec. 14	,,	98
Shooting Star ..	903	Baker	,,	Dec. 19	Boston	86
				1852		
Flying Cloud	1793	Creesy	,,	Jan. 6	New York	94
N. B. Palmer ..	1490	Low	,,	Jan. 9	,,	84
Eureka	1050	Canfield	,,	Feb. 9	,,	101
Mandarin ..	776	Stoddard	Woosung	Feb. 19	,,	109
Mermaid ..	—	——	Canton	Mar. 12	,,	87
Sea Witch ..	890	Frazer	,,	Mar. 21	,,	110
Raven	715	Henry	,,	Mar. 22	,,	107
Syren	1064	Silsbee	Manila	April 1	,,	103
R. B. Forbes ..	—	——	Canton	April 3	,,	104
Samuel Russell	940	——	,,	April 6	,,	95
Comet	1836	Gardner	,,	May 3	,,	99
Oriental	1003	Palmer	Woosung	Aug. 30	,,	107
Ariel	1340	Delano	,,	Sept. 2	,,	107
Honqua ..	706	——	,,	Sept. 14	,,	114
Swordfish ..	1036	Collins	Canton	Sept. 25	,,	90
Staghound ..	1535	Richardson	,,	Sept. 25	,,	95
Sea Serpent ..	1337	Howland	,,	Oct. 4	,,	88
Witchcraft ..	1310	Rogers	Woosung	Oct. 5	,,	117
Panama ..	670	——	,,	Oct. 26	,,	99
Shooting Star ..	903	Baker	,,	Nov. 13	,,	106
Flying Cloud ..	1793	Creesy	Canton	Dec. 1	,,	96
Atlanta ..	—	——	,,	Dec. 14	,,	84
White Squall ..	118	Lockwood	,,	Dec. 14	,,	103
Vancouver ..	—	——	Woosung	Dec. 26	,,	96
				1853		
Hurricane ..	1607	Very	Canton	Feb. 2	,,	99
Mandarin ..	776	Stoddard	Woosung	Feb. 19	,,	89
Sea Witch ..	890	Frazer	,,	Mar. 11	,,	106
John Wade ..	639	Willis	Canton	Mar. 19	,,	106
Southern Cross ..	950	Stevens	Manila	Mar. 19	,,	106
Raven	715	Henry	,,	Mar. 23	,,	102

The navigation laws, which were so protective as to give owners no inducement to make improvements, were repealed in 1849. Free trade was adopted in spite of the strenuous opposition of the ship-owner, and our foreign markets thrown open to the world. The American clippers at once took advantage of the chance to enter the British tea trade.

The "Oriental" Loads Tea for the British Market.

This vessel caused as great a sensation in England, as the *Rainbow* and *Sea Witch* had caused in America, being the first American ship to enter the West India Docks tea laden after the repeal of the navigation laws. In appearance she was a big edition of *Samuel Russell*. Her measurements were:—Length 185 feet, beam 36 feet, depth 21 feet. Her poop, which was stored with tea on the homeward run, was 25 feet long. She was a shortish ship with a moderate clipper bow, supporting very long bowsprit and jibbooms, and the heavy looking American stern. Like all American clippers she crossed skysail yards.

She left New York on the 14th September, 1849, on her maiden voyage, and reached Hong Kong by the Eastern Passage on 1st January, 1850, 109 days out. Sailed from China on the 30th, and ran home to New York in 81 days.

On her second voyage, she was commanded by Captain T. D. Palmer, his brother having retired from the sea. By this time Theodore Palmer had evidently learnt to drive a ship, for on this voyage he made both his own name and that of the ship.

On the passage out she left New York on 19th May, 1850; had very scant N.E. trades; crossed the line in long. 30½° W., 25 days out; log to the line 3904; best day's run 264 miles. Took 45 days to Cape Meridian. Best run from the line to the Cape Meridian 300 under double reefs part of the time, the breeze N.W. fresh. From lat. 42° S., long. 31° E. to long. 97° E., she averaged 264 miles a day, best day 302, worst 228, for 10 days. Passed St. Paul's Island 58 days out. Reached Anjer 29th July, 71 days out, and arrived Hong Kong 8th August, 81 days out, averaging 200 miles a day.

After this very fine performance three of the biggest tea firms in England gave their agents orders to secure her at any price, and she was at once chartered through Russell & Co. at £6 per ton of 40 cubic feet, whilst British ships lay waiting for tea at £3 10s. per ton of 50 cubic feet.

The *Oriental* loaded 1118 tons of tea, her freight amounting to £9600—almost three-quarters of her original cost.

She left Hong Kong on 28th August, had a strong S.W. monsoon, and yet beat down to Anjer in 21 days.

She signalled the Lizard 91 days out, and on 3rd December hauled into the West Indian Docks 97 days from Hong Kong. This wonderful sailing caused great excitement in English shipping circles, and all kinds of gloomy notices appeared in the papers, predicting the extinction of the British Mercantile Marine, etc. The Admiralty surveyors were even sent down to take off her lines whilst she lay in the dry dock at Blackwall.

The Aberdeen Clipper Model.

In the general despondency amongst shipowners, there seemed to be only one spark of hope, and that was in the Aberdeen model. On all hands it was admitted that no other type of ship could possibly rival the performance of the *Oriental.*

The Aberdeen model was devised by Alexander Hall, a ship-builder of that town, as far back as 1841. It was a very simple improvement, and merely consisted in carrying out the stem to the cutwater and giving a ship a long sharp bow instead of the old-fashioned apple-cheeks.

The first vessel built by Alexander Hall on this plan ran between Aberdeen and London, and soon made a great reputation by the rapidity of her passages.

Between 1841 and 1850 Messrs. Hall launched no fewer than 50 vessels, averaging 600 tons apiece, from the clipper schooner *Torrington* for the Chinese opium trade to the humble coaster. Their model was soon followed by other Aberdeen shipbuilders, the best known being Hood, the designer and builder of the famous clippers of the Aberdeen White Star Line, running to Australia and China.

Amongst these was the *John Bunyan* of 470 tons, whose 99 days coming home from Shanghai in the spring of 1850 was overshadowed by that of the *Oriental*; certainly the *John Bunyan* had a fair monsoon, but then Shanghai is some days further up the coast than Whampoa.

Two other Aberdeen clippers, the *Reindeer* of 328 tons, and the *Countess of Seafield* of 450 tons, made passages of under 110 days from Whampoa in *Oriental's* year.

The *Reindeer*, commanded by Captain Anthony Enright, was the first vessel to arrive with new teas in 1850, having made the voyage out to China and back in 7 months 28 days.

Her owners, Messrs. Fear & Vining, were so pleased that they presented Captain Enright with a chronometer.

"Stornoway" and "Chrysolite."

The immediate results of *Oriental's* success were further orders to Alexander Hall. Jardine, Matheson & Co. ordered the *Stornoway*, and Taylor & Potter, of Liverpool, ordered the *Chrysolite*.

These two ships have usually been spoken of as the first of the British tea clippers.

In design they had nothing in common with the great American clippers. The chief difference being the narrow beam of the Aberdeen ships, which is well shown in the following comparison of beams to length:—

Sea Witch, ...	5·01	*Stornoway*, ...	6·10
Surprise, ...	4·87	*Chrysolite*, ...	5·70
Oriental, ...	5·13		

This lack of beam made them far less powerful in heavy weather and they could not be driven in strong breezes like the Americans. Indeed, they had so little bearing forward that they went through the seas rather than over them, and gained a bad reputation for washing men off the jibboom when the huge jib of that day had to be handed.

At the same time they were undoubtedly faster and more weatherly than the Americans in moderate breezes.

The *Stornoway* was the first of the two to leave the ways. She was commanded by the crack racing skipper of his day, Robertson, who had come from the *John o' Gaunt*, which with the *Foam*, *Alexander Baring*, *Euphrates*, and *Monarch* had been the pick of the British tea fleet in the forties.

He took her out to Hong Kong in 102 days, and brought her home from Whampoa in 103 days.

"Chrysolite's" Maiden Voyage, 1851.

The *Chrysolite* also made a very good passage out to Hong Kong of 102 days, though she had a very bad time running her easting down. Captain Enright described his experience as follows:—

"In the Indian Ocean we had a gale from the west, and the tremendous seas that incessantly swept over the stern caused great injury to some of my men. My chief officer had his skull severely hurt. I did my best to plaster and bandage up the wound, but had scarcely done it before I was called to attend to one of the men who had his thigh broken. We shortened sail and lay to for a while, and I set the limb as well as I could.

"Altogether six of the men were severely injured by the heavy sea, and I know not how I escaped as all through that dark night I stood watching the steering. At daylight we sighted St. Paul's Island, and now saw that our misfortunes by stopping the ship, had probably been the means of saving our lives. Had we not laid to at the time we did, we should have been thrown on to the island, and in that dark night and furious sea not one of us could have escaped." *Chrysolite* was only 70 days to Anjer compared to the *Oriental's* 89 days, which was considered to be a record.

From Hong Kong the *Chrysolite* proceeded to Whampoa to discharge her cargo; and she barely had emptied her hold when the American clipper *Memnon* passed her going down the river with a cargo of new tea for London. A number of people happened to be on board the *Chrysolite* at the time, including her agents and the American consul, and bets were freely made by the champions of the two ships. Captain Enright, however, was one of the few tea clipper captains who never would bet on his vessel, and he remained true to his principles.

The *Chrysolite* took in a cargo at £4 4s. per ton and was ready for sea 3 days later. But she had no sooner finished loading when six of her men refused to go home in her.

Racing ships like the tea clippers were not always very popular with the old-fashioned merchant Jack owing to the constant sail shifting and freshening of the nip.

These men were given their discharge, but their departure left the ship with only sixteen men all told, and this scanty crew refused to weigh anchor without the complement being made up. At last Captain Enright quieted his small ship's company by promising to call at Hong Kong for more men.

The steamer came alongside to tow the clipper down the river late in the afternoon on 19th August.

The wind was right ahead for Hong Kong and the night pitch dark. As soon as the *Chrysolite* was under weigh Captain Enright called all hands aft and promised the men that if they would proceed on the passage without the additional hands, he would divide the wages of the six men he was short of amongst the remainder. He also reminded them that they were racing against an American ship, and that as Britishers they were bound to do their best to maintain the honour of their country.

The Chief officer, Partridge, at once tossed his hat in the air and shouted "I will go for one." The second mate followed his lead, and the crew, stirred by this example, at once gave in and the *Chrysolite* was soon on her way to Macao and the open sea.

On the twenty-third day out Pulo Glass was reached, and on that afternoon a large ship was sighted close to the Island of Banca and inside the coral reef.

Captain Enright, surmising that the ship ahead was the *Memnon*, determined to follow her inside the reef in spite of the great risk involved, the reef having to be crossed by a narrow channel through which a foul current was running very strongly. Captain Enright himself took the wheel. The passage was so narrow that when half way through the ship touched the reef on both bilges; however there was plenty of water under her keel and she scraped through without damage.

By 6 p.m. she was up alongside the ship ahead, which proved to be the *Memnon*. The latter's captain hailed Enright and asked him if he intended beating through Macclesfield Strait that night. Enright replied "Yes." Captain Gordon of the *Memnon* had then of course to do the same, though it led to his ship's destruction. Macclesfield Strait, which lies between Banca and Pulo Leat, was one of those favourite short cuts used by the racing tea skippers, but its navigation was very dangerous, especially at night, owing to a rock called "Discovery Rock," which lies in the centre of the channel and only has a few feet of water above it.

All that night the wind blew fiercely from S.W., and both ships beat to windward under all sail, the *Chrysolite* slowly gaining until she lost sight of her rival in a heavy squall. That squall was fatal to the *Memnon*, a beautiful ship of 1000 tons built by Smith & Dimon, of New York, in 1848 for Warren Delano.

Being forced into shallow water she let go both anchors, but in spite of them she began to bump on the hard coral bottom and a hole was soon stove in her. She was held hard and fast aft, though she had 8 fathoms of water under the bow, and it was soon found impossible to keep the water out of her. At daylight the ship was boarded by plundering Malays, and a number of the crew, who were mostly Malays and Portuguese shipped at Macao, joined the wreckers in stripping the ship.

Upon which, Captain Gordon, who was helpless in the face of their numbers, saw nothing for it but to abandon the ship, and he shortly set off for Gaspar Island with the three boats under the command of himself and his two mates. In his own boat he had his wife and her maid. Unfortunately they had not furnished the boats with enough water and owing to the heat suffered very much from thirst. They landed on the N.W. side of the island and luckily found water there. Captain Gordon at once established a lookout and also sent his bo's'n away to Singapore to procure help. He then left his people encamped on the island and went back to the wreck, but found the *Memnon* full of water and stripped of everything moveable.

After six days on Gaspar Island the shipwrecked crew were rescued by the barque *Jeremiah Garrett*, which in turn transferred them to the brig *J. M. Casselly*, bound for Singapore, where they arrived with little more than the clothes they stood up in.

Meanwhile the *Chrysolite* had made a fair passage to Mauritius, off which island she experienced the most terrific electric storm which lasted three days, during which time the sky was pitch black with the exception of the vivid lightning, the rain fell like a water-spout and the ship was shaken like a leaf by the tremendous rolling of the thunder.

Crossing the trades she made the following runs:—235, 264, 260, 180, 172, 225, 289, 290, 236, 230, 320, 260, 200, 212, and 268 miles—which was wonderful work for a vessel of her size. Going free under all sail she logged a steady 12½ to 13½ knots, and went up to 14 for limited periods; on the wind her best speed was at the rate of 10½ knots.

On 6th November in 8° N., the *Chrysolite* fell in with the homeward bound British frigate *Havannah* and the clipper ship *Fly*, both noted sailers. The three ships were in company for 18 days, the little white tea clipper maintaining a lead of from 2 to 5 miles. One day in calm weather Captain Enright actually dined aboard the *Havannah*, whose officers declared that the *Chrysolite* was the only vessel that had been able to hold them during the whole of their commission, and they were so convinced of her superiority that they gave him their letters to post on her arrival in England.

The three ships finally parted company 700 miles south of the Western Isles, the *Chrysolite* losing sight of the *Havannah*, which was then 10 miles on her lee beam, in a squall.

Passing the Azores the tea clipper had the misfortune to lose all three topgallant masts in a heavy puff. A ship named the *Adriatic* was in company at the time, and Captain Enright asked her in vain to lend him some spars; however he refitted as best he could and eventually had his revenge by arriving first by some days.

The *Chrysolite* entered the Princes Dock, Liverpool, on 1st December, 103 days out from Whampoa, after a remarkable maiden voyage, during which she had run 29,837 miles in 206 days.

Her owners were so pleased that they gave Enright a gratuity of £50. At the same time the ship's sailing so impressed brokers that she was able to command a guinea extra freight.

The Challenge of the American Navigation Club.

Though the American vessels sailing from China in the summer of 1851, with tea for London, did not make better passages, than either the *Stornoway* and *Chrysolite*, they still continued to command the best freights and to excite great admiration in the London docks.

Stirred by the success of their clippers, and at the same time not a little by the victory over English yachts gained by their wonderful schooner, the *America*, the American Navigation Club, which had no duplicate in England, issued a challenge to the British shipbuilders and owners, the conditions of which were that "two ships should be modelled, commanded and officered entirely by citizens of the United States and Great Britain respectively, and that they should sail with cargo on board from a port in England to a port in China and back to the English port, the prize for the winning vessel to be £10,000, and to be paid without regard to accidents or to any exceptions." There was also a stipulation made that the ships should not be under 800 tons or exceed 1200 tons American register. This challenge, I am sorry to say, was never taken up by British shipowners, notwithstanding the fact that the President of the American Navigation Club offered to raise the Prize money to £20,000 and to give the British vessel 14 days' start.

Dicky Green and the "Challenger."

It was Richard Green, of the famous Blackwall Line, who first put heart into the British shipowner. He had been one of the few people interested in shipping who had supported the Government in the repeal of the navigation laws. During 1851, at a large city dinner, he rose to speak after the secretary of the American Legation, who had made the usual complimentary international speech.

"We have heard," said Dicky Green, "a great deal tonight about the dismal prospects of British shipping, and we hear, too, from another quarter, a great deal about the British lion and the American eagle, and the way in which they are going to lie down together. Now I don't know anything about that, but this I do know that we, the British shipowners, have at last sat down to play a fair and open game with the Americans, and, by Jove, we'll trump them."

And he was as good as his word, for he immediately set about building a tea clipper, which he aptly named the *Challenger*. This vessel was sent out to China in time for the new teas of 1852.

The "Challenger" and the "Challenge."

The year 1852 was a remarkable one for the keenness of the racing between the tea clippers of the two nationalities, and *Challenger's* part in this rivalry was a very big one.

After loading tea at Shanghai she fell in with the famous *Challenge*, a vessel of nearly three times her size, at Anjer, the *Challenge* being homeward bound with Canton tea. The two vessels left Anjer on the same day, and when this was telegraphed home tremendous stakes were wagered as to which should be the first arrival in London; it was even rumoured that the loser was to be forfeited to the winner.

After a verysmart run the little *Challenger* just succeeded in beating the big *Challenge* into dock by two days, their actual times being:—

Challenger, Shanghai to London, 113 days.
Challenge, Whampao to London, 105 days.

Though beaten by two days in her time from Anjer, the *Challenge* actually made the shortest passage, but when one allows the difference in mileage between Shanghai and Whampoa it will be seen that there was very little to choose between the passages of the two vessels.

Curiously enough, as if in support of the forfeit rumour, all trace of the *Challenge* seems to be lost after this race. She went into the Blackwall Dock, where her lines were taken off, as those of the *Oriental* had been, and many people declared that she came out under the name of the *Result.*

The *Result* certainly was a big American-built ship, bought by Greens to take part in the booming Australian trade; but she was a smaller ship altogether from the *Challenge.*

A more likely story was that Messrs. Green bought her with the money won over the race. This, however, does not throw any light on the after life of the *Challenge*, for which later vessels, named after her, have often been mistaken.

One of these was a 1200-ton ship, built in Quebec in 1863. Another was a clipper called the *Golden Age*, which in her last days was renamed *Challenge.* There was some excuse for confusing this vessel with the original *Challenge*, as she was exceedingly fast, with the tall reputation of having run 22 knots an hour, and at the same time was terribly strained from hard driving. She was, however, a flush-decked ship with an inner rail a few feet in from the covering board, and only resembled the real *Challenge* in the length of her mainyard.

Naval Science of July, 1873, gives the following:—10th July, 1868, *Challenge* left Shanghai, arrived London 131 days out. 6th August, 1869, *Challenge* left Shanghai, arrived London 148 days out.

These passages may possibly be those of the real ship, but it is more likely that they refer to a British-built clipper barque of some 500 tons. So I fear I must leave the fate of the notorious *Challenge* unsolved, in the hope that one of my readers may be able to shed light upon it.*

"Witch of the Wave's" Passage Home in 1852.

The best passage in 1852 between China and London was made by the Salem clipper *Witch of the Wave*, Captain Millett. But it is slightly discounted by the fact that she sailed in the height of the N.E. monsoon, which took her from Whampoa to Java Head in 7 days 12 hours, a record time. Loading 19,000 chests of the choicest tea, she sailed from the Canton River on 5th January. On 13th January she had cleared Sunda Strait, and she was 29 days between Java Head and the Cape, her best run crossing the trades being 338 miles. The last part of the passage was rather interfered with by easterly winds, and she was four days beating up channel. She took her pilot off Dungeness on the 4th of April, and docked 90 days out from China.

Whilst she was lying in the River she attracted great attention, and her bows and general appearance were considered to resemble those of the famous *America*.

Her wonderful passage caused American ships to be more popular than ever amongst shippers in China, the *Chrysolite*, *Stornoway*, and *Challenger* being the only British ships which were taken up to load new teas.

Race between "Stornoway" and "Chrysolite."

Stornoway and *Chrysolite* both loaded tea at Whampoa. Whilst waiting to load, Captain Enright made the acquaintance of a very interesting personality, namely, General Garibaldi, who was then in command of a Chilean barque of 450 tons. General Garibaldi

* Since writing the above, a letter from the last owner of the *Challenge* has come into my hands. *See* Appendix J.—B.L.

had been brought up as a sailor, and it was to the sea that he generally returned after the success or failure of his guerilla campaigns in the cause of liberty. When Enright met him he was still mourning the death of his wife, Anetta, who had died in his arms through exposure in an open boat on the Adriatic, when they were fleeing from the French. And he told Enright with tears of grief and rage how, when he had buried her in the sand at the back of a farmhouse the French had exhumed her body thinking to find his hidden treasure. Strangely enough, when a boy, Garibaldi had sailed in a vessel belonging to Enright's uncle and, with this as a bond between them, the two men struck up a great friendship.

Getting away together on the 9th July, *Stornoway* and *Chrysolite* were in company for 21 days down the China Sea, neither ship proving able to give the other the slip.

Of this race to Anjer, Captain Enright remarked in his personal log: "I remained on deck night and day in a bamboo chair, made fast to the skylight. I nodded occasionally, but only when I dared allow myself a few moment's rest from my ceaseless watching of the wind and course. However, all my efforts to get ahead were unavailing, the *Stornoway* being the larger vessel and better sailer."

It was a ding-dong race in spite of this admission of Enright's, the two vessels being in company for no less than 45 days altogether, but in the last lap the best winds fell to *Chrysolite*, and she arrived first, docking in Liverpool on 22nd October, 104 days out. *Stornoway* reached the Downs three days later, 107 days out.

The *Chrysolite* was the first ship in with the new tea of the season, and Captain Enright again received a gratuity from his owners.

So great was the demand for her tea that the *Liverpool Mercury* records: "*Chrysolite* was docked on Saturday morning at 9 o'clock, and before night a considerable portion of the cargo was landed, weighed, duty paid, and about 100 chests of tea were on their way to distant parts of the Kingdom, and a quantity of it was in the hands of retail dealers in the town, so that no doubt it was actually upon the tea tables of some of the people of Liverpool the same night—an instance of despatch unparallelled in this or any other port of the Kingdom."

Best Tea Passages of 1852.

Two other American ships besides the *Witch of the Waves* and *Challenge* made good passages this year. The *Surprise*, Captain Dumaresq, left Whampoa on 19th July, and arrived at Deal on 2nd November, 106 days out, whilst the *Nightingale* came home from Shanghai in 110 days, this being the best passage from that port.

It seems that there was a lot of wrangling this year as to which ship could claim the best passage home in the S.W. monsoon. And so convinced were the owners of the *Nightingale* of their vessel's superiority, that they challenged the world for a £10,000 China race, but though their challenge was directed against American clippers as much as English, like that of the American Navigation Club, it met with no response.

"Cairngorm."

Noticing the way in which their little clippers were handicapped when in competition with the Americans, which were usually about double their tonnage, Messrs. Hall, of Aberdeen, in the winter of 1852, laid down a 1000-ton clipper with finer lines and stronger scantling than anything they had built up to that date. They also used iron for the main deck beams and half of the hold beams, which gave more room for stowage.

This vessel, which was called the *Cairngorm*, was bought by Jardine, Matheson & Co., who placed Captain Robertson of the *Stornoway* in command.

The *Cairngorm* has been given the credit of being the first British clipper which really could rival, if not excel, the wonderful American ships. Nautical writers of the day declared that not only did she beat them on time, but that owing to the superior strength of her build she delivered her tea in much better condition.

Tea Passages of 1853.

The *Cairngorm*, on her first voyage, loaded tea in Shanghai, and made the best homeward passage of the year from that port, her time being under 110 days.

She was closely followed by the little *Challenger*, which arrived in London on 3rd December, 110 days out, having handsomely beaten the American clippers *Nightingale* and *John Bertram* and won £4000 in stakes.

The Americans, however, were revenged by the little Baltimore wonder, *Architect*, which arrived in the Downs from Whampoa, 107 days out, and had sold her cargo before the arrival of the first Britisher from that port, amongst the vanquished being the *Chrysolite* and *Stornoway*, and such noted fast ships as the *Hero of the Nile*. This performance gained *Architect* £2 per ton extra freight on her next tea passage.

"Lord of the Isles" and "Northfleet."

In the year 1853 two very celebrated ships were built for the China trade. The first of these two was the *Lord of the Isles*, one of those vessels which was in advance of her times.

She was built by Messrs. Scott, of Greenock, a firm which had not previously tried their hands at a tea clipper. In a great many ways she was a radical departure from the accepted type as shown by the great Aberdeen builders.

In the first place she was constructed entirely of iron, and secondly, she had more beams to length than any vessel launched within ten years of her appearance, the proportion being 6·4. She was also so fine in the ends that she was nicknamed "The Diving Bell," and Maxton, her celebrated skipper, used to say that she dived in at one side of a sea and out at the other.

On her first voyage she went out to Sydney in 70 days, and in 1858 she came home from Shanghai in the record time of 89 days. However, in spite of her success and the fact that she "delivered her cargo without one spot of damage," iron ships were never popular in the tea trade, and it was not until the seventies that any more iron tea clippers were built.

Lord of the Isles had a short life. On 24th July, 1862, when bound to Hong Kong from Greenock, she caught fire in lat. 12° 13′ N., long. 115° 50′ E., in consequence, it was supposed, of the "Spontaneous combustions of some bales of felt placed in juxtaposition with bundles of railroad iron in the lowerhold." Captain Davies, his crew and passengers, in all thirty souls, managed to make Macao in the boats, after being twice boarded by pirates.

The *Northfleet* was a very different vessel, one of Duncan Dunbar's superb frigate-built ships.

She was constructed at Northfleet on the Thames and must have been an extraordinarily fast vessel for her type. In the years 1857 and 1858 she made two splendid passages out to Hong Kong from Woolwich of 88 days and 88 days 7 hours; and the *Shipping Gazette* gives her times for a passage home in 1857, which, if correct, constitute a wonderful record. She apparently left Hong Kong on 8th August, passed Anjer 7th September, and arrived Plymouth on 29th October, 82 days out.

Northfleet's end was one of the most tragic in the annals of marine disasters. She was lying at anchor off Dungeness, outward bound in 1873 for Tasmania with emigrants. Suddenly at 10.30 p.m. when most of her passengers and crew were asleep below, she was cut down to the water's edge by an unknown steamer, which backed out and left her to her fate.

As the *Northfleet* began to sink, there was a most terrible panic amongst the emigrants, who were mostly labourers without any knowledge of ships or the sea. These men rushed the boats in spite of the revolvers of Captain Knowles and his officers, with the result that 293 souls, including many women perished.

The ship took half an hour to go down, and but for the panic all hands might have been saved. Captain Knowles went down at his post, but his wife was saved, and was given a pension from the Civil List in recognition of her husband's bravery.

The steamer that did the damage turned out to be the Spanish steamer *Murillo*, but at the inquiry into his conduct her captain escaped punishment on the assertion that he had no idea that his ship had done any damage.

Tea Passages of 1854.

The chief international race of 1854 was that between the *Chrysolite* and *Celestial*.

The former sailed from Foochow and the latter from Whampoa on 14th July, and *Chrysolite* arrived at Deal 108 days out, one day ahead of her rival.

The *Challenger* and *Stornoway* left Shanghai together, were together in Sunda Strait, and the former arrived at Gravesend on the 4th December, three days ahead of *Stornoway.*

Cairngorm again made the best passage from Shanghai, but a new clipper, the *Crest of the Wave*, sailing on 20th October, was only a few days behind her in time. This year Hall, of Aberdeen, built the little *Vision* for James Beazley. She measured 563 tons, and was only 6 months 16 days at sea on her maiden voyage from Liverpool to China and back.

"Nightingale's" Passage in 1855.

Nightingale was the last of the American clippers to distinguish herself in the English tea trade. In 1855 she ran to London from Shanghai in 91 days, her best run being 336 miles.

From this date the Americans, who two or three years back were a common sight in the London docks, gradually fell out of the English trade, and America Square near the Minories, which had been the headquarters of their skippers, slowly grew deserted.

British Clippers of the late Fifties.

At the same time with new captains and water-soaked hulls the careers of the first British clippers *Stornoway* and *Chrysolite* were about over. After four splendid voyages in the *Chrysolite*, Captain Enright left her to take command of the great American-built Black Baller *Lightning*, whilst Robertson had shifted from *Stornoway* to the *Cairngorm.*

Meanwhile new clippers kept coming out, which stayed in the front of the battle for a few seasons and then dropped into the ranks of the has-beens.

Of such were the Sunderland-built *Crest of the Wave* and *Spirit of the Age*, the successors of Sunderland's frigate-built veterans *John Temperly*, *Sir Harry Parkes*, *Dartmouth*, *Harkaway*, etc.

These two vessels like the Aberdeen White Star ships went out to Australia before crossing to China to load tea, the passage out to Australia in the early days of the gold discovery being as keenly contested as the homeward one from China.

Regarding this, an interesting story is told of *Spirit of the Age*. It happened that some ten or a dozen prime British shellbacks were looking round the West India Docks in search of a ship, when they came upon a crack Yankee clipper displaying a large notice board in her rigging, on which the following words were painted: "This ship will be in Australia before any other."

Whilst they were gazing at this early specimen of American advertising the captain of the *Spirit of the Age* came up to them and pointing at the Yankee's skysign, asked with a show of passion:

"How can you stand that, and you Britishers?"

"Well, it's a bit cheeky maybe," replied one of the Jacks. "but how can we help it?"

"That's what I've come to you for," returned the captain of the *Spirit of the Age*, eagerly, "you are all looking for berths, aren't you?"

"Yes," answered the men's spokesman.

"Well", went on the enterprising Sunderland skipper, "I have all my crew on board already—and a good crowd too; but come with me as a sort of extra watch and I'll see if we can't bustle the bounce out of that Yankee."

And the men did ship with him, with the result that the *Spirit of the Age*, after being driven night and day, made the passage to Sydney in 73 days, and beat the boastful American, who was about three times her size, by a day and a half.

The Yankee's signboard was a specimen of the shrewd business qualities shown by American captains.

A Yankee Captain's Cuteness.

There were many other stories of Yankee cuteness in the days of America's great mercantile rivalry with Great Britain, of which, perhaps, the following is the best.

One year when freights at Whampoa had dropped and only vessels with noted reputations for speed were getting good rates, an American ship arrived in the port. This vessel was far from being a fast sailor, nevertheless her captain went boldly up to the skipper of the fastest British tea clipper in port and challenged him to a race home for £500 a side.

This was at once accepted; and the crafty Yankee then saw to it that the facts of the wager were made public. On the shippers hearing of it, they immediately concluded that his vessel possessed a turn of speed which had been a carefully kept secret. They thereupon engaged her at £2 per ton more freight than she would have been offered if her real sailing powers had been known.

The two ships sailed for England, and the British crack had no difficulty in beating her opponent by a fortnight. But after paying over his stake of £500, the American captain was still nearly £1000 to the good, owing to the extra £2 on his freight, which but for his crafty method of gaining his ship a reputation for speed would never have been offered him.

"Kate Carnie" and "Fiery Cross."

In 1855 the celebrated firm of Robert Steele & Co., built their first tea clipper, the *Kate Carnie,* a little ship of under 600 tons. That she proved a success goes without saying, but she is chiefly remembered as being the first of a long list of beautiful ships launched from Steele's yard at Greenock, amongst which were such cracks as *Serica, Taeping, Ariel, Sir Lancelot, Titania* and *Lahloo.*

In the same year Chaloner of Liverpool launched the first *Fiery Cross,* which in the hands of a very exceptional skipper named Dallas made some very fine passages. Unfortunately Dallas left her in 1869 to superintend the building of a ship, afterwards celebrated as the *Fiery Cross* No. 2, and on her passage out that autumn under a new captain *Fiery Cross* No. 1 was wrecked.

"Robin Hood" and "Friar Tuck."

The great rivals of *Kate Carnie* and *Fiery Cross* (No. 1) were the Aberdeen -built *Robin Hood* and *Friar Tuck,* vessels which were considered the *Sir Lancelots* of their day, although it was a short one.

The *Friar Tuck* was wrecked in St. Mary's Roads, Scilly Isles, when homeward bound in the sixties; her cargo floated ashore, and the prime China tea which was retailed through Cornwall by the Cornish wreckers will long be remembered there. Her figure-head is at Tresco Abbey, Scillies.

Tea Passages of 1856.

The year 1856 is notable as being the first year in which the

£1 per ton premium on the freight was offered for the first tea ship to arrive in the London docks.

The chief American clippers taking part in the racing were the *Maury*, *Ringleader*, *Celestial*, *Comet* and *Quickstep*.

The *Vision*, Whampoa to Liverpool, was the first ship home, and *Cairngorm* the second.

The *Lord of the Isles* was the first ship away from Foochow. She left on 9th June, followed by *Maury* on 13th June.

The *Maury* was a beautiful clipper barque of some 600 tons, built by Rossevelt & Joyce for A. A. Low & Brother. She made a splendid race of it with *Lord of the Isles*, the two vessels arriving in the Downs on the same day, 15th October, *Lord of the Isles* being 128 days out and *Maury* 124 days. Off Gravesend *Maury* was leading by 10 minutes, but coming up the river Captain Maxton had the smartest tug, and thus managed to dock first and win the premium.

Chrysolite, under Captain Jock McLeland, left Whampoa on the same day that *Lord of the Isles* left Foochow, but she did not have a fortunate passage. In the China Sea she carried away all three topmasts. Upon which the crew refused duty, and brought a beef bone aft to show the skipper that before clearing away the wreck they required better grub or more of it. Not gaining any concession from the hardened old man, one of the men threw the beef bone at him, whereupon Captain Jock whipped out a derringer and winged the man, who dropped to the deck yelling blue murder. In telling the yarn afterwards the mate remarked that he had never seen a deck so quickly cleared. It appeared that Captain McLeland never came on deck without his derringer. With this kind of friction going on between captain and crew it is not surprising that *Chrysolite* took 144 days to get home.

The best passage from Foochow was made by *Fiery Cross*, which, leaving on 4th September, was under the 100 days to London.

Spirit of the Age did the best passage from Whampoa. Leaving 20th August, she arrived in the Downs on 18th November, exactly 100 days out. *Ringleader* came next with 109 days, leaving 14th July, and reaching London on 31st October.

Of the Shanghai ships, the *Challenger* left on 8th September and came home in 129 days. This was not one of her best, but all the passages from Shanghai were long ones that year. *Challenger's* average for eight passages from Shanghai was 110 days, the shortest run being 105. All these were made under Captain Killick, who eventually left her and founded the firm of Killick & Martin, and owned the well-known clippers *Kaisow*, *Omba*, *Wylo*, *Osaka*, and *Lothair*.

Tea Passages of 1857.

Ship	Port from	Date Sailed	Arrived	Date Arrived	Days out
Crest of the Wave	Foochow	May 25	London	Sept. 28	126
Maury	,,	July 3	London	Oct. 17	106
Cairngorm ..	Hong Kong	July 10	Deal	Oct. 30	112
Northfleet ..	,,	Aug. 8	Plymouth	Oct. 29	82
Challenger ..	Shanghai	Aug. 5	London	Dec. 1	118
Robin Hood ..	Foochow	Aug. 6	London	Nov. 30	116
Spirit of the Age	,,	Aug. 8	Liverpool	Dec. 1	115
Fiery Cross ..	,,	Anjer Aug. 29	Off Dartmouth	Nov. 11	fr. Anjer 74
Celestial.. ..	Shanghai	Aug. 23	London	Jan. 11	141
Lord of the Isles	,,	Aug. 25	London	Jan. 13	141

I have been unable to find *Fiery Cross's* sailing date, but she probably left Foochow about the beginning of August and thus came home in less than 110 days, a very fine performance.

I have already mentioned *Northfleet's* record. It is as given in the *Shipping Gazette*, which records her passing Anjer on 7th September, so that would only give her 52 days from Sunda Strait home, which is manifestly impossible, and yet I can find no evidence of it being a misprint.

Celestial and *Lord of the Isles* must have sailed a close race, but they evidently had poor winds on the China coast, as I find that *Lord of the Isles* was 45 days to Anjer, which she passed on 9th October.

Fiery Cross took the premium for the first ship in, but this

Tea Passages of 1858.

Ship	Port from	Date Sailed	Passed Anjer	Arrived	Date Arrived	Days out
Fiery Cross ..	Foochow	June 27	—	London	Oct. 20	115
Chrysolite ..	Foochow	July 8	—	,,	Nov. 26	141
Northfleet ..	Hong Kong	July 22	Aug. 24	Plymouth	Nov. 25	126
Kate Carnie ..	Foochow	Aug. 2	—	London	Dec. 2	122
Stornoway ..	Foochow	Sept. 6	—	,,	Jan. 21	137
Robin Hood ..	Foochow	Sept. 8	Oct. 7	,,	Dec. 17	100
Challenger ..	Shanghai	Sept. 18	Oct. 22	,,	Jan. 11	115
Cairngorm ..	Whampoa	Nov. 6	Banca St Nov. 18	,,	Feb. 6	92
Lammermuir ..	Whampoa	Nov. 8	—	,,	Feb. 9	93
Lord of the Isles	Shanghai	Nov. 29	—	Dover	Feb. 26	89

year is remarkable for the wonderful times of the late starters, which, with an N.E. monsoon and good winds in the Atlantic, had rather an advantage over the first ships.

The abstract of *Lord of the Isles*' 89-day run was as follows:—

Left Shanghai	November 29
Rounded the Cape	January 14
Passed St. Helena	January 23
Crossed the line	February 2 in long. 23° 40′ W.
Passed the Lizard	February 25
Arrived off Dover	February 26

She carried 1030 tons of tea, and it was given out that she had averaged 320 knots for five consecutive days crossing the trades in the Indian Ocean.

Race between "Cairngorm" and "Lammermuir."

Owing to the war with China the Canton River was closed to European commerce in 1857 and 1858; but the operations of the British fleet under Admiral Sir Michael Seymour opened Whampoa to trade in the Autumn of 1858, upon which five ships rushed up to load tea at that port.

These were *Cairngorm*, *Lammermuir*, *Chieftain*, *Warrior Queen*, and *Morning Star*. Of these the *Cairngorm* was the only out-and-out clipper. *Lammermuir*, which had been built by Pile of Sunderland for John Willis in 1856, was not a sharp-ended ship, but she was a good all round sailer, and particularly fast in light winds. *Chieftain* was a Jersey built ship and considered fast, but neither the *Warrior Queen* nor *Morning Star* had any reputation for speed.

However, as it was the first tea that had been loaded in the Canton River for some time, the merchants were very eager to get it on the home market, and so promised a large bonus to the captains and officers of the first ship to arrive in London. It was considered 100 to 1 on the *Cairngorm*, which got away 2 days before the *Lammermuir* and the other ships.

Lammermuir was commanded by Captain Andrew Shewan, senior, who afterwards had the *Norman Court*, whilst Captain Ryrie, who afterwards had the *Flying Spur* had succeeded Robertson at the helm of *Cairngorm*.

With light weather in the China Sea *Cairngorm*, like many another very sharp ship, was not at her best—the Aberdeen clippers excelled chiefly in strong, whole-sail breezes—and the *Lammermuir* caught her in the Java Sea when eight days out. The wind was very light and just suited the Sunderland ship, which soon forged up alongside her famous rival, upon which Captain Shewan lowered a boat and went on board the *Cairngorm* to have a chat. And whilst his captain was aboard *Lammermuir's* antagonist, the mate, a man named Moore, who afterwards held command of several of Willis's clippers including the *Cutty Sark*, tacked the *Lammermuir* across the other's bows and so sailed round her, much to Captain Ryrie's disgust, which was expressed in some very forcible language.

The two vessels did not part until they felt the S.E. trade outside Java Head, but then *Cairngorm* ran away from *Lammermuir* in the fresh breeze. Nevertheless the latter docked only three days behind the Aberdeen crack after a splendid passage. *Lammermuir's* performance was considered so meritorious that the brokers presented the same bonus to her captain and officers as that which they had given to *Cairngorm* for being the first arrival.

Of the other three ships, the *Chieftain* was only two or three days behind *Lammermuir*, but the *Warrior Queen* and *Morning Star* were nowhere.

Lammermuir, the heroine of this race, eventuallv left her bones on the Amherst Reef, Macclesfield Channel, Gaspar Straits, and her mainmast was for many years a good guiding mark.

"Ellen Rodger" and "Ziba."

In 1858 two very fast little clippers were built, the *Ellen Rodger* from Steele's yard and the barque *Ziba* from Hall's. Their day, however, was only a short one, as they were soon outclassed by the improved clippers which appeared at the beginning of the sixties.

"Chaa-sze."

Before leaving the fifties for the sixties and the great days of the tea races, I will mention one more of the earlier Aberdeen clippers, the *Chaa-sze*, which followed *Ziba* off the stocks, and was the first tea ship designed by Rennie. This little ship was of more than usual interest. Measuring 556 tons, and built by Hall, she was laid down for a steam whaler, and made as strong as it was possible, being of teak throughout. Inside her teak frames, which were 4 to 6 feet apart, she was diagonally planked. This with 3-inch outside planking bound her up tightly with no less than 9 inches of teak. Indeed, she was so tightly bound that, on being put into the tea trade, she had several of her lower deck beams made to unship in order to give her more play. This idea was taken perhaps from the trick of the old slavers, which, when hard pressed by a cruiser, would saw through their deck beams in order to improve their sailing, and, indeed, often attributed their escapes to this cause.

The *Chaa-sze* made some very good passages from Whampoa in the early sixties.

On one occasion she was in company, off Mauritius, with a Sunderland clipper barque named the *Chanticleer*, a predecessor of *Maitland* and *Undine*. The two ships had been in company for four days with light variable winds, both being bound to China. At last a steady breeze arrived, and *Chaa-sze* at once began to show her heels and drop the other. Upon which Geordie, the skipper of the *Chanticleer*, remarked resignedly to his mate: "There she goes. They have unscrewed the beams, and we shan't see her again."

And no more they did.

PART II

"The Zenith of the Tea Clipper Racing" 1859 *to* 1872.

The Builders and Designers of the Famous Tea Clippers.

WITH the advent of the sixties the British tea clippers came to their perfection; their star rose to its zenith, and for a year or two after the opening of the Suez Canal still shone brightly—then came steam and the eclipse.

With the retreat of the Americans, it was left for the British designers and builders to fight out the battle for supremacy amongst themselves. And the fight between Aberdeen and the Clyde grew to be just as keen as that between Great Britain and America.

The Clyde possessed two great builders in Steele and Connel, who, until the launch of the famous *Thermopylae*, may be said to have thrust the productions of Hall and Hood into the shade. And of these two, the firm of Steele held the palm. Built by Robert Steele and designed by his brother William Steele, such vessels as *Taeping*, *Serica*, *Ariel*, *Sir Lancelot*, *Titania*, and *Lahloo*, carried all before them in the tea races, besides being the most beautiful and yacht-like merchantmen that ever sailed the seas.

Their most notable rivals from Connel's yard were *Taitsing*, *Spindrift*, and *Windhover*, but Connel must also receive credit for Skinner's *Castles* which were very smart little ships, but, being much smaller than the cracks, did not race in the first flight with the new teas.

Hall's best known ships of the same date were *Flying Spur*, *Black Prince*, and *Yangtze*, whilst Hood built *Jerusalem*, *Thyatira* and *Thermopylae* for Thompson's Aberdeen White Star Line. Of these three the *Thermopylae* was the only vessel which regularly loaded the first teas of the season.

A few other builders entered the arena from time to time. The well-known Pile of Sunderland failed to hold his own. His *Maitland*, of which so much was expected, was a failure, and *Undine*, though she made a fair average of passages, was not quite equal to the Steele cracks. Green of Blackwall built *Highflyer* and put her into the China trade for a couple of voyages, but even under Anthony Enright she did not prove fast enough. Chaloner of Liverpool produced one masterpiece, the famous *Fiery Cross*; Laurie of Glasgow built *Leander*; Stephen of Glasgow launched some half-dozen ships for the trade, including *Forward Ho*; whilst the wonderful *Cutty Sark* was built by Scott & Linton of Dumbarton.

As a rule these famous clippers were designed in the drawing lofts of their builders; in fact, there were only two outside designers of any note, Bernard Waymouth, Secretary of Lloyd's Register, and Rennie.

Waymouth was responsible for the lines of the *Leander* and *Thermopylae*, whilst Rennie designed *Fiery Cross*, *Black Prince*, *Norman Court*, and *John R. Worcester*.

The Beauty of Steele's Creations.

Though there was no such thing as an ugly tea clipper, Steele was, without a doubt, the designer of the most beautiful little ships that ever floated. Like his modern confrere, Fife, he could not produce an ugly boat. The lines of his vessel never failed to please the eye; their sweetness and beauty satisfied that artistic sense in a sailor, which, though always present, can hardly be described in words. Suffice it to say that there was not a curve or line or angle in a boat such as the *Ariel* or *Sir Lancelot*, which did not carry out the idea of perfect proportion and balance. And it is just this balance in design which gives a ship merit in the eyes of sailors. Steele's gracefully curving cutwaters and neatly rounded sterns fitted each other to perfection. His vessels never gave one the impression, as some boats do, that the bows of a ship had been joined on to the stern of a scow and *vice versa*. And it was this absolute sweetness to the eye which gave the Steele clippers a look of delicate, almost fragile, beauty, and distinguished them from their rivals. The Clyde clippers, also, were noted for a yacht-like finish: all their woodwork on deck or below was of the finest teak or mahogany, so beautifully fashioned as to bear comparison with the work of a first-class cabinet-maker, whilst bulwark rails, stanchions, skylights, capstans, and binnacles shone with more brass-work than is ever found in a modern yacht.

Pride of the Clyde Shipwrights.

The building of tea clippers on the Clyde laid the foundations of that river's supremacy in ship construction. Many are the stories related of the pride of Greenock and Glasgow shipwrights at this period. How they had their foot-rule pockets made shallow on purpose to show this emblem of their trade. How they would swagger into the barbers' shops and demand to be shaved before the ordinary customer because they were shipwrights.

They even took up oarsmanship, and a crew of Clyde shipwrights calling themselves the "Cartsdyke Worthies," actually succeeded in winning the four-oared Championship of Britain at the Thames National Regatta in 1871.

Craze for Neatness aloft in Aberdeen Ships.

The Aberdeen ships, though not so expensively finished as those of the Clyde with regard to deck fittings, were celebrated for their smart look aloft. Indeed they carried this smartness aloft to excess, with the result that, owing to the smallness of their blocks they were heavy workers; and it was not until freights had begun to fall that large blocks and small ropes replaced small blocks and large ropes. The *Black Prince*, a noted ship for small blocks, required two men to stick out her fore-sheet, and kept all her heavy braces and sheets rove through the light weather of the China Seas. And Captain Shewan, who was an officer on both ships, tells me that the *Norman Court*, though 100 tons bigger, handled more easily with 14 A.B.'s than the *Black Prince* with 22. The Aberdeen ships, though perhaps the worst offenders with regard to small blocks and heavy ropes, were not the only ones. The fault of mistaking weight for strength seems to be a characteristic of the British nation. As a proof of this, so heavy was the usual stunsail gear of British ships that it was a common sight to see a Frenchman or Italian set a whole suit of stunsails, whilst a Britisher was setting a fore topmast stunsail.

Sail Plans of the Crack Clippers.

At the beginning of the sixties most of the clippers had Cunningham's patent fitted to their single topsails, and it was not until 1865 that a double-topsail was seen in a tea ship. But in the autumn of that year *Ariel* came out with double-topsail yards on all three masts, followed a month later by *Sir Lancelot* with double on the fore and main, but retaining the single yard on the mizen.

The double topsail had one disadvantage, it was not so effective in light winds as the single topsail owing to the splitting of the sail. As a proof of this fact, Captain Keay wrote to me, *apropos* of his celebrated race up channel with *Taeping* in 1866, that when the wind slackened *Taeping* had slightly the better of it, whereas, as soon as it freshened, *Ariel* went ahead; and he accounted for this difference in their sailing by the fact that *Taeping's* single topsails were more effective in light weather than his double ones. Double topsails also necessitated the slacking up of the lee topmast rigging when by the wind in order to allow the lower yard to brace up well. When racing it was customary on the tea clippers to lace the foot of the upper topsails to the lower topsail yards. The vessels had their full complement of stunsails (or studding sails to spell them correctly) from royal stunsails down to save-alls and watersails which were set under the lower stunsails.

In most of the tea clippers the topgallant stunsails were set from the deck. In the favourable trades, such as when running from Anjer to Mauritius or rolling down to St. Helena, a staysail was laced on as a wing outside the lower stunsail. Captain Keay of *Ariel* even went so far as to lace two staysails together, thus making a square sail to go outside the lower stunsail. A small set and a large set of stunsails were carried on these well-found clippers.

Passarees were boomed outboard some 30 feet at the fore, and when before the wind the foresail was set as flat as possible with its clews hauled well out on the passaree booms, whilst the clews of the mainsail were carried aft.

Staysails were bent on every stay, including that from the main skysail masthead. It was the universal practice amongst the later tea clippers to haul down their staysails when close-hauled or turning to windward. The epoch-making *Falcon* started this fashion, and claimed that it enabled her to lay half-a-point nearer the wind than her rivals.

Some of the earlier clippers, owing to their sharp bows, had a lot of rake in their spars, and often carried a great deal of weather helm. To overcome this a jib-o-jib or jib-topsail was set well up on the fore-royal or fore-topgallant stay, and a sail called a Jamie Green made of No. 4 canvas, and cut as a main-topgallant stunsail with three feet more hoist, was set along the bowsprit and jibboom under the headsails. This, a very favourite sail on the tea clippers, was filled mostly from the fool-wind of the jib, but pulled hard, every inch telling both in light and moderate winds.

It was always set on a wind, being carried even when making short tacks by the smart ships, and the foc's'le-head men attended to it when going about.

It will perhaps interest modern sailors to hear how this sail was set. The sail was run out and in along rope travellers rove between the end of the jibboom and the cat-heads. The halliards went to the jibboom end, and the tack to the lower end of the martingale, and the inner or sheet clew of the sail was flattened or eased off by means of a pendant from the fore-rigging and a whip to the foc's'le-head.

A ring-tail was, of course, set outside the spanker with a water-sail under it or else a save-all under the spanker. A bonnet, also, was generally laced under the foresail. The crossjack of a tea clipper was seldom set on a wind as it was difficult to make it stand, overlapped as it was by the mainsail and spanker.

One of the largest sails set when racing was the main topmast staysail, which stretched the whole length of the stay. Most of the earlier clippers carried a huge jib, but in the later ships this very unhandy sail, which caused the death of many a man in the first Aberdeen flyers, was split into the modern style inner and outer jib.

The Steele clippers were noted for their fairy-like main-sky-sails, but the Aberdeen boats and most of the others carried nothing above their royals, and relied more on spread than hoist. A few ships, however, sent up temporary skysail yards in favourable weather, and the *Maitland* alone set moonsails above standing skysail yards.

These moonsails of hers were, however, a standing joke amongst the tea fleet, and were compared to pocket-handerchiefs.

Deck Plans.

The deck plans were generally alike—a short topgallant foc's'le termed a monkey foc's'le; a small midship-house; the boats carried on low skids between the main and mizen masts just forward of the mizen rigging; and a raised quarter-deck, flush with the main rail and extending a few feet forward of the mizen-mast.

Dead-Rise and Ballasting.

The midship section of the *Sir Lancelot* shows the amount of dead-rise usual in Steele's productions. Steele, like the American, Donald Mackay, in his later ships, believed in a full midship section and fine ends, but some clipper ship designers cut their ships away almost like yachts. All the tea clippers required a deal of ballast, and besides some 100 tons of permanent iron kentledge stowed under the skin in the limbers, they took in over 200 tons of washed shingle before loading tea.

It required very little to alter the trim an inch or two. *Ariel*, in the 1866 race, used a shifting box 12 feet by 3½ by 2, made of 3-inch deals, and filled with spare kedges, anchor stocks, and coal, to trim to windward. And the moving of salt provisions from the fore peak, or putting the cables aft in the sail room, was quite enough to put the stern down an inch or two.

When levelling the ballast before loading tea it was usual to trim about 3 feet by the stern, so that when loaded the ship often drew 4 or 5 inches more aft than forward.

To get her correct trim was as important in these sensitive tea clippers as it is in a modern yacht, and a half inch one way or the other often made all the difference in a ship's sailing.

Sheer.

Like the modern racing yacht, the tea clippers had just the right proportion of sheer, and in this respect came half way between the Blackwall frigates, which had absolutely none, and the Yankee clippers, which in many cases carried sheer to excess.

Rigidity of Build.

When the rivalry in the tea trade between America and the United Kingdom was at its height the British vessels were noted for the good condition in which they delivered their cargoes, whereas in American ships the tea was often injured owing to the lightness of their construction and the soft woods employed, which allowed the water to drain into the hull when under the strain of a heavy press of sail.

The British composite-built clippers of the sixties were, however, so beautifully built as to be as tight as a bottle. Such ships as the *Ariel* required a 10 minute's spell at the pumps every 24 hours of the first week outward bound, after which they had taken up, were perfectly tight, and required no more pumping for the voyage.

Towards the end of the tea races, however, a theory grew up in clipper ship circles that the old rigidity of build was a mistake, that a vessel sailed better if allowed some play in hull as well as gear; and this theory, with which many a racing yachtsmen will agree, was carried out in some of the later vessels. For instance, it is reported that *Thermopylae's* deck seams would open up when heavily pressed.

It is quite possible, however, to get strength without rigidity, and this is proved by the length of life of some of the hardest driven. The *Cutty Sark* voyaged regularly between Lisbon, Rio, and New Orleans, up to 1922, and both *Thermopylae* and *Titania* survived the beginning of the twentieth century.

Speed of Tea Clippers Compared with the Black Ballers, Yankee Clippers, and Later Iron Clippers.

In comparing the speed of different vessels it is necessary to take a number of factors into account, such as trim, quality of cargo, condition of ship's bottom, character of captain, and strength of crew, before being able to come to a fair judgement.

In the tea clippers and Australian Black Ballers all these factors were excellent. The Yankee clippers usually had to contend with weak or mutinous crews, whilst the iron clippers often had to put up with weak crews and foul bottoms. But allowance, especially in heavy weather, must also be made for tonnage, and the little tea clippers were little more than half the size of the others.

Thus it would not be reasonable to expect them to rival such vessels as the Black Baller's *Lightning* and *James Baines* or the Yankee clippers *Flying Cloud* and *Sovereign of the Seas* when running the easting down. The Black Baller, owing to her size and height out of the water (emigrants were a light, easy cargo, giving a high freeboard) could run before the westerlies with dry decks and skysails set, when a tea clipper, with her narrow beam and low freeboard, would only be burying herself if pressed or half-becalmed under the lee of each roller if snugged down to the lower canvas.

But in the light weather of the tropics and more especially in the baffling airs of the doldrums, the little tea clippers could sail 2 feet to a Black Baller's one. I have taken out the times between the two tropics from the logs of various ships and find that the tea clippers were usually five or six days faster than either the Black Ballers, Yankee clippers or the iron clippers.

On 20th September, 1855, the Black Baller *Lightning* crossed the tropic of Cancer and was 25 days to Capricorn, whilst *Ariel* crossed Cancer on 22nd September, 1865, and only took 18 days between the tropics. I have chosen this instance, as they crossed the tropics at the same season of the year and experienced pretty much the same weather.

Ariel's best time between the tropics was in November, 1866, as days; *Thermopylae's* 12 days in November, 1866; *Lightning's* 16 days in February, 1855; and *Patriarch's* (to give an example of an iron clipper) 15 days in June, 1883. I have picked these four ships as they were probably as fast as, if not faster than, any others of their type.

In doldrum weather such vessels as *Ariel, Thermopylae, Sir Lancelot* and *Titania* possessed the power of ghosting along 4 or 5 knots when there was scarce a ripple on the water and when a Black Baller or Californian flyer would barely have had steerage way.

Weatherliness of the Tea Clippers.

At the same time they went to windward in a wonderful manner. A good instance of this was shown during *Thermopylae's* first voyage, when leaving Melbourne for Newcastle, N.S.W.

"Thermopylae" beating to Windward.

The tug towed her 2 miles below the Gellibrand Lightship; she then cast off, sail was set and she stood over towards the St. Kilda Bank; stayed on the port tack headed for Point Cook; went round again and then fetched the head of the South Channel. The pilot, who took her down, was amazed; and told the tug master that he had taken many ships to sea under similar conditions, but the weatherly qualities of the *Thermopylae* eclipsed anything he had ever seen. No vessel, he declared, had ever made the South Channel in three tacks before with the wind from the same quarter.

Weatherliness of "Sir Lancelot" and "Ariel."

The sister ships *Sir Lancelot* and *Ariel* were specially noted for the way in which they could beat dead to windward in a strong breeze.

The following entries in Captain Keay's abstracts testify to *Ariel's* powers:—

—. "Ship goes 12 knots on a bowline quite easy."

—. "Ship going close-hauled 8 or 10 knots, pitching much, lee side of deck constantly flooded, water coming over bow and lee quarter close aft. Distance in 24 hours 222 miles against such a sea."

—. "Fresh gales, severe squalls, very high turbulent sea, she behaves splendidly, going 11 knots, against such a sea. Distance by observation from 9 p.m. to noon 174 miles."

I have no instances of the tea clippers meeting larger clippers in fresh winds, but in light weather the *Flying Spur* once passed the *Lightning* very easily; she likewise went by *Sabraon* and *Tantallon Castle*. *Ariel* also overhauled the latter vessel with great ease, the wind being light and ahead and stunsails in.

Best Day's Run of a British Tea Clipper.

The best day's work of the tea clippers was generally done in smooth water with a strong whole-sail wind about 2 points or so abaft the beam. The *Cutty Sark* holds the record with 363 knots, done more than once. On one occasion she did 362 and 363 knots on two consecutive days. On another she made 182 knots in 12 hours.

The *Thermopylae's* best was 358 knots, made when running her easting down in 44° S., 68° E.

The *Ariel's* best was 340 and *Sir Lancelot's* 336.

Speed of the Crack Tea Clippers Compared.

The *Falcon's* best point was to windward, where she showed a great superiority over her predecessors. Captain Keay, who commanded *Falcon* and *Ariel* in turn, says that *Ariel* was a knot faster all round than *Falcon.*

Titania with more beam was stiffer and not so ticklish to handle as the two sister ships *Ariel* and *Sir Lancelot. Spindrift* was very fast off the wind but the Steele clippers had slightly the best of her on a wind.

Fiery Cross, Taeping, Scrica and *Lahloo* with their single topsails, were at their best in light breezes. *Kaisow,* a very narrow ship, was not as fast as her contemporaries except in light winds. *Forward Ho* and *Windhover* were good wholesome all-round ships, and very fast when hard sailed, which was not often. *Leander* was considered by many to be as fast as *Thermopylae,* but she suffered from bad captains. *Lothair* was one of the fastest of the lot in light and moderate breezes, but had not the power to stand driving in heavy weather. *Norman Court* could outweather and outsail the fleet on a wind but was not so fast running. *Thermopylae* and *Cutty Sark,* being larger and more powerful, stood driving in heavy weather better than the graceful Steele flyers and had much the best of it when running their easters down. In hard breezes *Cutty Sark* was the fastest ship of the fleet, but in light weather *Thermopylae* and the Steele cracks could beat her.

Yet taking them all round there was very little difference in speed between the best known of the clippers, and in the racing one can safely say that their captains had as much or more to do with their success or failure than the ships themselves.

I have collected, where I could, instances of ships in company, which will show at once the level sailing of the first rankers.

Ariel and *Spindrift* were over a week in company in the China Seas when homeward bound in 1868, and *Ariel* only succeeded in getting through Anjer Strait ahead by a daring piece of navigation on the part of her commander.

On another occasion *Ariel* was six days in company with *Fiery Cross*, the wind being mostly ahead and from a fresh breeze to light airs, and Captain Keay told me that in all that time he only gained about a mile or a couple of points to weather.

Flying Spur and *Sir Lancelot* in 1867 were ten days in company running down to St. Helena. Again *Flying Spur* was in company with *Taeping* for seven days.

Sir Lancelot and *Norman Court* were a week in company going down the China Sea homeward bound in 1874. In the same year *Norman Court* and *Kaisow* were in sight of each other most of the way between Beachy Head and the line.

In 1872, when racing home, *Thermopylae* and *Cutty Sark* were within a few hours of each other from Shanghai right down to the Cape, where, *Cutty Sark*, when leading, had the misfortune to lose her rudder.

These few instances will show how narrow was the margin in speed between the clippers.

The Handling of a Tea Clipper.

The handling of a tea clipper was a ticklish business, and the captain who went into the tea races after being used to slower and less sensitive craft often found himself all at sea and made a bad mess of it at first.

A case in point was the dismasting of the *Titania.* In clipper ships it was bad practice to put your helm up in a squall, though the Board of Trade only recognised that manoeuvre when one was passing the examiners.

Experienced tea-ship captains invariably gave strict orders to an officer, who had just come out of a non-clipper, never to keep away in a squall, but to luff and shake the squall out of her, though the officer had, of course, to be careful not to get his ship aback, and there was also the danger of splitting sails.

The danger of putting the helm up in a sensitive and heavily-sparred clipper was this. As the wind freed the ship gathered more way, and, her yards being more fore and aft owing to her long lower masts than those of other ships, the sails got the full weight of the squall abeam. If the ship was the least bit tender, or it was an extra heavy puff, she would put her rail under so far that the helm lost its power over her. Then, probably, the halliards would be let fly, but, owing to the angle at which the ship was heeled, the yards would not come down, which meant that something had to go.

In *Titania's* case, she encountered a fierce squall just north of the Cape Verds. Her captain, Bobby Deas, who had come from a wagon called the *Reigate*, ordered the helm to be put up. Even so, if he had been in time to get the ship off the wind before the weight of the squall struck her, all would have been well; but he was too late. The squall caught her square on the beam. She went right over until her fairleads were in the water. The topsail yards stuck at the mastheads, and away went the foremast, jibboom, main topmast, and mizen topgallant mast.

The *Titania*, *Ariel*, and *Sir Lancelot* were ships that required very careful handling and wanted knowing, but once a captain got the hang of them they would do anything for him but speak.

These three ships were very fine aft, with a counter like a yacht, which had a nasty habit in bad weather of dishing up the seas. This fineness aft also caused them to be troublesome boats to put about in a rough sea, as they fetched sternway so quickly, and, of course, then took heavy water aboard aft. Thus it was customary to wear them round when there was a nasty sea running.

No greater proof of the way these Steele clippers were cut away aft can be given than the story of how Captain England backed the *Titania* up the Shanghai River. In turning up the river he found that she stirred the mud up every time she came about and was very slow on stays. So on one board, instead of staying, he threw everything aback, brought her stern up to the wind, and sailed her across backwards; and this he continued, making one tack bow first and the other stern first.

The fact is that the Steele clippers were a wee bit too fine aft. On the other hand, the early Aberdeen clippers did not have enough bearing forward, with the result that they were terribly wet in anything of a head sea.

Most of the tea clippers were inclined to be on the tender side. Some of them, indeed, were so overhatted as to be dangerously crank, but such ships were never in the first flight. But when cleverly handled no square-rigged ship that ever sailed the seas was as handy and willing as these beautiful little tea ships.

The Owners.

Amongst the tea-ship owners there were many men of the good old-fashioned type, who loved their ships and took more interest in them than they did in their balance sheets. Though keen enough business men, they had that pride in their ships which insisted that everything, down to the smallest detail, should be of the very best, Such owners trusted their captains and gave them a free hand, and were as liberal and generous in their treatment of their employees as they were regarding the outfit and upkeep of their ships. Their vessels never lacked paint or rope any more than their captains lacked a bonus or their crews a sufficiency of good provisions.

Many of them were old sea captains, who had either set up for themselves or else, owing to their great experience and distingusished careers, had obtained partnerships in good firms. Of such were Maxton, who commanded *Lord of the Isles* and *Falcon*; Rodger, who commanded *Kate Carnie*; and Killick, who commanded the *Challenger*. And most of the other owners had at one time or another made voyages on their own ships, and knew them as a sailor knows a ship.

These men, as I have said, loved their ships, and the loss of a ship meant far more to them than a mere pocket loss of cash. A case in point was the wreck of the *Spindrift* on Dungeness, which broke old Findlay's heart and resulted in his mind giving way.

Probably one of the best known of the old-fashioned type of shipowners was John Willis, Captain John, as he was called in shipping circles, who, with his white hat, was almost as familiar a landmark as the dock capstan on the pierhead of the West India Docks. No ship of his ever departed or arrived without his personal farewell or greeting at the dock gates.

The following extracts from a friendly article in *Fairplay* will give a good idea of this fine old shipowner:—

"John, better known as Captain John, commanded one of his father's ships before he was well out of his teens, and his contemporaries will tell you no smarter man ever trod a quarter-deck. Nothing but the best and plenty of it will do for John when his ships are in question. Once when one of his ships had run off her class, and it was found that it would cost more to replace her than she was worth, he sold her to a ship breaker at a low price in order that she should be broken up, having refused a higher price for her to work. 'If,' he said, 'she is not good enough for me she ought not to go to sea again.' If I were to say that John's temper is exactly that of a lamb I am afraid I should be rough on the lambs. He is as hard as iron and as straight as the day. He calls a spade a spade and if he does not like a man he calls him a scoundrel. That is his word. He does not mean much by it and he applies it indiscriminately to anything that offends him. John says what he thinks, but no one is much the worse for it. He is rich and a bachelor, a favourite with the ladies. With whatever faults he has he is a fine specimen of the old-fashioned, high-class shipowner."

After his trouble with the underwriters over *Black Adder's* misfortunes John Willis never insured his ships. He was also peculiar in never allowing his captains to have any interest in them.

The Captains.

No man had more to do with the reputation of a ship than her captain. In the China trade daring, enterprise, and endurance were the *Sine qua nons* of a successful skipper. And many a speedy ship, as we shall see, was never given a chance of doing herself justice, owing to her misfortunes in the way of captains. First-class men were so scarce that I can barely scrape up a dozen worthy of remembrance.

There were many safe, steady goers, but these were not the passage makers. It required dash and steadiness, daring and prudence to make a crack racing skipper, and these are not attributes of character which are often found in conjunction. A born racing skipper has always been as rare as a born cavalry leader, and those in command of the tea ships proved no exception to this rule. Most men were either too cautious or too reckless—added to which the China Coast was very wet (sailor's parlance) in those days, and a drunken captain was too often the explanation of a fine ship's non-success.

However there were a few men, who held the necessary qualities of a tea-ship commander, whose endurance equalled their energy, whose daring was tempered by a good judgment, whose business capabilities were on a par with their seamanship, and whose nerves were of cast iron. These men could easily be picked out of the ruck, for their ships were invariably in the front of the battle. Amongst the best known were Robinson of *Sir Lancelot*, Keay of *Ariel*, McKinnon of *Taeping*, Kemball of *Thermopylae*, Andrew Shewan of *Norman Court*, Burgoyne of *Titania*, John Smith of *Lahloo*, and Orchard of *Lothair*.

There are many ways of making a passage, and as many of sailing a ship. Some captains invariably made good tracks, others did not bother about mileage as long as they could keep their ships moving, others again prided themselves on their daring navigation in cutting corners and dashing through narrow channels at night.

The clippers, like thoroughbred horses, responded to the master touch like things of life; Robinson, for instance, was said to be worth an extra half-knot an hour on any ship; this could only be done by the most sleepless vigilance.

Thus the strain of a three months' race was naturally tremendous. Some captains only went below to change their clothes or take a bath; others used the settee in the chart-room or even a deck-chair as a bed. This was the habit of old Captain Robertson, of the *Cairngorm*, who during the homeward run never turned in but dosed with one eye open in a deck-chair on the poop.

Captain Keay, of *Ariel*, writing of his passage down the China Coast, remarks: "My habit during those weeks was never to undress except for my morning bath, and that often took the place of sleep. The naps I had were of the briefest and were mostly on deck."

Many a man broke down after a few years of it, but the giants, such as Keay or Robinson, went on and on without a rest, and, still more wonderful, with hardly a serious accident.

Ruses used by the Captains against one another.

The excitement of the racing was, of course, doubled when the ships were in company. Some captains had a strong dislike to sailing in company, as it increased the tension; thus Captain

Care of the *Lord Macaulay* invariably had a man stationed on his fore royal yard, whose timely warning enabled him to keep the horizon between himself and a rival.

Speaking of this habit of Captain Care's reminds me of a trick he once played on the *Elizabeth Nicholson*.

The *Lord Macaulay* was approaching a narrow passage between two islands in the Java Seas, and it was getting on for sundown when her rival, a new ship, the *Elizabeth Nicholson*, with a captain who was also new to the China Seas, was sighted astern bringing up a breeze.

Captain Care who knew the passage well, and, who had meant to go through it during the night, determined to take advantage of the other captain's inexperience in an effort to shake him off.

He began to shorten sail as if he meant to bring up for the night, and was at once gratified by seeing the guileless captain of the *Elizabeth Nicholson* prepare to follow suit. In order to give the other ship time, Captain Care pretended to miss stays, then as soon as the *Nicholson* was within hearing distance, he sang out loudly: "Stand by and let go the anchor." It was then just on dusk. Care waited until he heard the plunge of the *Nicholson's* anchor and the roar of the chain through the hawse-pipe, upon which he at once filled away again on the *Lord Macaulay*, and slipping through the channel during the night, was thus enabled to gain a lead of 70 miles.

He did not sight the *Nicholson* again until he was off the Scillies, when she was seen away to the norrard. Care managed to get into the Lizard, then the nor'-west wind coming away strong, he boomed up channel and arrived nearly a week ahead. He was specially pleased at this victory, as the skipper of the *Elizabeth Nicholson* had treated the idea of the *Lord Macaulay* being able to beat his new ship with scorn, and was so strait-laced into the bargain that he refused even to bet the proverbial hat on the result.

It was in quite a different fashion that Captain Robinson of the *Sir Lancelot* fooled Captain Innes of the *Spindrift* in the 1869 race. The *Spindrift* had sailed from Foochow on the 4th July, 13 days before *Sir Lancelot*; nevertheless Robinson managed to overhaul Innes off the Cape, and one fine clear morning the *Spindrift* sighted a ship on her starboard beam, which signalled the number of the *City of Dunedin.* On his arrival Captain Innes reported speaking the *City of Dunedin* on 31st August off the Cape, little knowing that the vessel was in reality his rival, the *Sir Lancelot,* which had already arrived in the Thames, five days ahead of him.

On another occasion Captain Keay in the *Ariel* got the better of *Spindrift.* This was in 1868. The two vessels had travelled nearly the whole length of the China Coast in company until one evening found them almost becalmed off the West Coast of Borneo. Both ships stood in to get the land breeze until at 8 p.m. they were so close on top of the land that *Spindrift* went about and stood out to sea again. But Captain Keay, putting out all his lights, held on and with his lead going crept nearer and nearer the shore. At midnight he had 9 fathoms. Then came the first puff of the land breeze and he immediately hove round on the port tack. The next three casts of the lead gave 5, 4½ and 4 fathoms, so he was obliged to keep off a bit, but all the time the breeze was freshening, and as he stood away on his course he was rewarded for his daring by a last glimpse of *Spindrift's* port light as she lay becalmed in the offing.

On moonlight nights daring captains often stole a march on their rivals by cutting through Stolzes Channel in Gaspar Straits or taking the Allas Passage out of the Banda Sea instead of the Ombai Passage. I have already described Captain Enright's daring navigation inside the Coral Reef off Banca Island, the night when the American clipper *Memnon* got ashore.

The Pride of Captains in their Ships.

The following instance will give some idea of the pride these captains had in their beautiful ships.

Captain Stainton Clarke who commanded the four-masted barque *Loch Carron* for so many years, served his time and finally gained his first command in Skinner's Castles.

Whilst he was serving as an officer on board the *Edinburgh Castle,* he was sent aboard the *Titania* one evening with a message for her captain, the notorious Dandy Dunn, whose nickname was the result of predilection for frockcoats and lavendar kid gloves.

As soon as Clarke had delivered himself of his message, Captain Dunn, asked:

"Is this the first time you have been aboard the *Titania*?"

"Yes, sir," replied young Clarke.

Upon which Dandy Dunn called out impressively to his first officer:

"Mister, take a lamp and show Mr. Clarke over the ship."

Tea Clipper Crews.

The crews of the tea clippers would make a modern ship-master's mouth water. Britishers to a man, they were prime seamen and entered into the racing with all the zest of thorough sportsmen. Many are the stories of their keenness on the homeward run.

Thermopylae's crews are stated to have spread their blankets in the rigging as an auxiliary to her sails.

In the great race of 1866, the crews of *Serica* and *Fiery Cross* backed a month's pay against each other.

"Often," relates Captain Care," have I seen the hands racing aloft in nothing but their shirts at the cry of 'All hands reef topsails.'" And this, mind you, not in the tropics but off the Stormy Cape or in the cold North Atlantic weather.

Captain Keay used to time his crew putting the *Ariel* about. In ten minutes from the calling of all hands, the *Ariel* was round, yards trimmed, ropes coiled down clear for stays, bowlines hauled, the tacks down and the watch sent below.

And Captain Shewan of *Norman Court* declared: "With all hands going about, we would have the ropes coiled up in ten minutes from the ready about order."

The Shanghai pilot once timed the *Norman Court* getting underway and shore that her anchor was lifted and sail made in twenty minutes. She was an easy working ship and her crew were accustomed to walk her topsail yards to the masthead in smooth water.

The following instance of smartness in repairing damage I take from Captain Keay's abstract of *Ariel's* maiden passage:—

"Saturday, 7th October, 1865.—7.30 a.m., main topgallant mast broke short off by the cap and at royal masthead—in three pieces, and the rigging broke the after topmast crosstree. Called all hands to send down the wreck. Brought topgallant yard to collar topmast stay and sent down the sail and other yards, etc., The starboard watch got the other mast hoisted out of mainmast, which was no easy job, and when got on deck found the heel so splintered and chafed that had we been where another could be got, it would not have been sent aloft. Bolted the splintered heel together and got it ready for going aloft.

"At 1 p.m., watch busy getting rigging cleared and mast ready (not work for all hands about the mast), found that hoop and iron grummet at royal masthead had been lost overboard.

"6 p.m., commenced to heave the mast up.

"9.30 p.m., fidded the mast; got stays and backstays set up by midnight.

"Sunday, 8th October, 1865.—Proceeding to get main top-gallant yard across and sail bent and set; royal yard up and sail set.

"8.30 a.m., got finished, cleared the decks up."

I could quote many other instances of this sort to show what clipper ship crews could do, but this will easily be seen when I come to describe the racing.

There has been a good deal said about the double crews of the tea clippers. As a matter of fact, they were by no means over-manned, especially when freights began to fall; and, when one remembers the crews of sixty to eighty men carried by the little 1000-ton Blackwall frigates, one is almost inclined to think that the tea ships had barely sufficient men.

In 1860, when freights were at their height, *Lord Macaulay* had a crew of 40 all told. And *Flying Spur's* complement consisted of twenty able seamen, two ordinary seamen, two midshipmen, bo's'n, sailmaker, carpenter, joiner, butcher, cook, two stewards, three officers, and captains—thirty-six all told. *Ariel* and *Sir Lancelot* carried captain, two mates, bo's'n, sailmaker, carpenter, cook, steward, and twenty-four able seamen—a total of thirty-two. This was for the passage home, when two extra A.B.s were generally shipped at Hong Kong.

But it must be remembered that every man was an A.B. in the fullest sense of the term. With torn sails constantly under repair it was necessary that every man should know how to use a palm. Such entries as the following were continually on the work slate:—"Watch side seaming and repairing torn sails."

Undoubtedly the crew of a tea clipper had very little rest when racing. This, however, was made up for by the excitement. The tension of the racing was never off, and spread to all hands, who caught the exhilaration of it and became animated with a fine *esprit de corps*, such as is almost as dead as the dodo in these modern days of machinery and self-interest, trade unions and ship managers.

"Thermopylae's" Cock.

No story shows the pride of the crews in their ships better than the theft of *Thermopylae's* cock. When she arrived in Foochow on her maiden voyage, after two record passages, she surprised the other ships at the Pagoda Anchorage by exhibiting a gilded cock of victory at her main truck. This was too much for the crews of the other vessels, which already had tea races to their credit, whilst *Thermopylae* still had hers to win. The story goes that a sailor on the *Taeping* jumped overboard and swam across to the *Thermopylae* whilst her officers and crew were having a grand spread below in honour perhaps, of her captain's birthday. Climbing the cable, he got aboard unseen, and soon removed the cock from its proud position, then shinning to the deck swam safely back to his own ship with the emblem of victory in his arms. When the *Thermopylae's* crew discovered their loss the fat was in the fire, and words, if not blows, occurred when any of the rival crews met. Indeed, the incident caused a great deal of trouble, and nearly led to a lawsuit. Captain Allan, of the Aberdeen liner *Marathon*, who was mate of the *Thermopylae* at the time, declares that the plot to carry off the cock was hatched by the officers and crews of all the other clippers. But, be this as it may, the removal of the proud emblem was carried out as above.

Thermopylae never recovered her golden cock, but she soon replaced it with another, which she carried proudly at her main truck for the rest of her existence.

Outward and Intermediate Passages.

The voyages of the tea clippers, though barely a year in length, showed a remarkable mileage. The outward passage was either to Hong Kong, Shanghai, or Melbourne. The outward cargoes were heavy ones, consisting generally of Manchester bales and lead.

Thermopylae, of course, held the record to Melbourne with her famous 60 days passage. *Ariel's* 80 days to Hong Kong was also a wonderful performance, accomplished as it was by the long eastern route and against the monsoon. *Leander* is credited with the record of 96 days to Shanghai, closely followed by *Cutty Sark* with 98 days in 1870-71.

Between their arrival in China and the time for loading the first teas the clippers traded up and down the coast, sometimes as far north as Japan, at others round to Singapore and Rangoon, but, as a rule, carrying rice from Saigon, Bangkok, and other rice ports to Hong Kong.

The following epitomes will give some idea of the mileage covered.—

Ariel in 1867.

London to Hong Kong, Hong Kong to Yokohama, Yokohama to Hong Kong, Hong Kong to Saigon, Saigon to Hong Kong, Hong Kong to Foochow, Foochow home.

Ariel in 1868.

London to Shanghai, Shanghai to Hong Kong, Hong Kong to Saigon, Saigon to Hong Kong, Hong Kong to Saigon, Saigon to Hong Kong, Hong Kong to Foochow, Foochow home.

Sir Lancelot in 1869.

London to Hong Kong, Hong Kong to Bangkok, Bangkok to Hong Kong, Hong Kong to Saigon, Saigon to Yokohama, Yokohama to Foochow, Foochow home.

Normancourt in 1872.

London to Shanghai, Shanghai to Swatow, Swatow, to Kobe, Kobe to Hong Kong, Hong Kong to Whampoa, Whampoa to Macao, Macao home.

Life on the Coast.

It was a pleasant life on the coast, and the tea clippers were, as a rule, happy ships. Though the work was strenuous and the navigation often perilous there was always a spice of excitement to keep monotony away.

As Joseph Conrad rightly remarks: "The China Seas, north and south, are narrow seas. They are seas full of every-day, eloquent facts, such as islands, sandbanks, reefs, swift, and changeable currents—tangled facts that nevertheless speak to a seaman in clear and definite language."

In those days the charts were by no means as correct and complete as they are now. I only have to quote a passage or two from Captain Keay's abstracts to show this:—

"1st December, 1865.—by p.m. sights found that Ambla is on chart about 11 miles too far west, if Savu N.E. Point and Ombai East Point are laid down right (confirmed following voyages).

"8th December, 1865.—Keil Island does not exist as placed on Imray's chart of 1856. Angour Island pretty right. Saw no appearance of shoal on its west side. On one occasion worked short tacks to southard along west side of Pellew Reef for about 40 miles north of Angour. Saw nothing like Keil Island. West edge of reef very distinct, nearly awash, with many heads of rock showing 10 or 12 feet above water.

"24th February, 1867.—P.M., stood in towards Onval Point, some 3 or 4 miles to leeward of spot marked Portsmouth Breakers, plied on and off directly over the spot without seeing anything of them from fore topsail yard."

Captain Shewan, senior, using Dutch charts for Gaspar Straits, found that the coast of Borneo about Tanjong Datoo was 10 miles out in latitude.

And no doubt many of the innumerable wrecks are to be accounted for by incorrect charts. I find that, of the tea clippers, the following were lost in Chinese waters:—*Fiery Cross* (No. 1), *Loochoo*, *King Arthur*, *Japan*, *Childers*, *Young Lochinvar*, *Taewan*, *Guinevere*, (No. 1), *Taeping*, *Serica*, *Lahloo*, *Ellen Rodger*, *Black Prince*, and *Chinaman*.

The Pilots on the Coast—Chinese and European.

The ships were further handicapped by the unreliability of the pilots. It was considered a most risky thing to take a Chinese pilot. They knew the waters well enough, but were generally in the pay of the pirates, or even coast fishermen, and thus rarely missed an opportunity of putting the ships in their charge ashore or wrecking them on some uncharted rock, which they purposely kept secret for such occasions.

Such a rock was the pinnacle rock at the mouth of the Min River, where the charts gave 15 fathoms. This rock was struck by the *Norman Court* in 1878. The European pilots declared that it must have been a sunken wreck. However, when the clipper was docked in Shanghai, oysters were found sticking in her bottom. Captain Delano, of the Yankee clipper *Golden State*, also stated that he had had a shoal cast about the same place. But it was not until two years later, when the *Benjamin Aymar* had stuck on an uncharted rock close by and remained there, that the pilots began to believe in the *Norman Court's* rock. Then H.M. gunboat *Moorhen* was sent down, and found a pinnacle rock, only 9 feet below low water springs, right on Captain Shewan's bearings.

Curiously enough, just before the *Norman Court* had discovered this uncharted rock with her keel, she had successfully employed a Chinese pilot. Coming down from Shanghai for Foochow in thick N.E. monsoon weather, Captain Shewan, on hauling in for the regular channel, found himself to leeward of the White Dogs. He picked up a Chinese pilot at daybreak. It would have taken a day beating up for the usual Channel, and when the pilot said: "Suppose you like, I can take the ship in as we go, I savvy plenty water, can do all right," Captain Shewan agreed to risk it, knowing that the Chinaman was licensed by the consul. And the pilot took him through a short cut into the Min River without mishap.

Owing to the heavy drinking on the coast, the European pilots were often not much more reliable than the Chinese. Perhaps the best known was old Hughie Sutherland of Shanghai, a Caithness man.

Many are the stories told of this character. He was a notoriously hard drinker, so you can imagine the surprise of Charlie M'Caslin the Californian skipper of the Shanghai towboat, the *Orphan*, when, on boarding an inbound ship, he once found old Hughie calmly drinking milk. However, his only comment was, "Too late, Hughie, too late."

Shortly after this incident old Hughie was hailed before the magistrates, who would not adjudicate because they said he was too drunk.

"You had better decide," says Hughie, "for I'll be drunker tomorrow."

However, Hughie Sutherland was a good enough pilot when sober, though a daring one.

Once when the *Norman Court* was leaving Shanghai for Foochow to load poles back, he came aboard declaring that he had turned total abstainer, meant to take a holiday and would go the trip with Captain Shewan. It was the end of September. On leaving Woosung, the *Norman Court* picked up a nice N.E. breeze.

Off Cape Yangtze, two steam coasters were sighted ahead bearing away south for the Bonham Pass.

Old Hughie looked hard at them for a moment and then turning to Captain Shewan, said:

"What do you say? Shall we follow those steamers?"

(The orthodox course for a sailing ship was round the Saddles.)

"A bit risky, isn't it?" replied the captain.

"Look here!" said Hughie, seeing the other in doubt. "This nor'-easter came away this morning, didn't it?"

"Yes."

"Well, you know they always last 24 hours."

"Yes."

"Well, by that time, we shall be through the Narrows, will we not?"

"Yes."

"Then, why not? Anyway we have always got the anchors."

"Quite right," agreed Shewan, "let her go."

So they squared the mainyard and went flying after the steamers. Old Hughie was right; the breeze held; the *Norman Court* soon overhauled the coasters and dropped them astern; and she got through safely before dark, thus saving a day on the passage down to Foochow.

That trip the *Norman Court* was only 23 days from Shanghai to Foochow and back. On the return passage poor old Hughie fell ill and died and they buried him at sea off the Hushan Islands.

Captain Shewan had left his wife at Shanghai, and his critics always put down this fast trip to her, but in reality it was chiefly owing to the daring pilotage of old Hughie. He always went into the chains himself with the lead, and if it was possible to save a tack he would do so.

Chinese Pirates.

There were other dangers in the China Seas, to be reckoned with besides in different pilots, rocks, shoals and treacherous currents. Pirates swarmed along the coast. For protection against these, every tea clipper was provided with an armoury of muskets, pistols and cutlasses besides two cannons, which were capable of more than ornamental or saluting duties. And they had special magazines for powder, ball and grape shot, small arms, ammunition, etc.

Regarding the use of the cannon for saluting purposes, an amusing incident occurred in Shanghai in 1868.

Several of the tea clippers were lying moored in the river. discharging, etc., before proceeding to Foochow, when the *Leander*, having finished, unmoored and towed to sea. Her departure was signalised by a general salute from the other clippers present.

And it so happened that just as the *Argonaut* fired one of her guns from the starboard side of the poop, a Chinese man-of-war junk happened to be sailing past.

The wad, which had been made purposely hard of old rope in order to raise a loud report, went right through the junk's mainsail and landed on the quarter of another ship near by, knocking away some of the gilt carving on her stern. The men on the junk fell flat on the deck with fright when the gun went off, and it knocked a hole in her sail large enough to drive a wagon through.

Though I can find no instance of these cannons ever having been used for self-defence, their mere presence often had the desired effect of keeping the pirates off.

"Lord Macaulay" and the Pirate Lorchas.

This is proved by an adventure of the *Lord Macaulay*. It happened that in order to avoid a typhoon, Captain Care put back into a bay near Hong Kong. On running into the bay he found it full of pirate lorchas. However, he determined to face the pirates rather than the typhoon, as the lesser of two evils, so he let go his anchor and putting on a bold front lowered away his boat and rowed off to each junk in turn on a tour of inspection; he pretended that he thought they were fishermen and asked each one if they had any fish, then satisfied in his mind as to their real character, he returned to his own vessel. All through that night he stood his crew to arms but whether they were influenced by his cannon, were kept quiet by the nearness of the typhoon, or thought he was not worth the trouble of attacking, the pirates kept off, and the following morning he slipped away.

The Pirates and "Ariel's" Sampan in Hong Kong Harbour.

The neighbourhood of Hong Kong was especially infested with the most daring pirates. They even ventured into the harbour itself on cutting out expeditions.

An incident of this sort I find in Captain Keay's abstract log of *Ariel's* first voyage:—

"20th April, 1866.—Last night at 11.30 p.m., whilst below reading, I heard screaming and loud cries from Ahoy and others in the sampan astern. I ran on deck, found the watchman Williams looking at the sampan and told him to haul the boat up that we might get down to help them as they were being attacked by men in a long 'pull-away' boat. There were seventeen of them, twelve got into our sampan and five remained in the 'pull-away' boat. They cut the rope so that the boats drifted away and hoisted the sampan's sails to run her out of the harbour, but before getting the rudder clear and steerage way, they got under the bows of a barque next astern of us.

"Directly it was seen that we could not get into the sampan, I hailed the *White Adder* to send men in his sampan; seven men jumped in with cutlasses but before getting there the pirates had cleared out, leaving a chopper and large knife in their hurry to be off. I also hailed ships astern to stop them if their boats were out, but none succeeded.

"Ahoy's mother had her hand cut, Ahoy was gripped by the throat, and was suddenly released by striking backward with his elbow; probably the blow took fatal effect as the man was silenced, and Ahoy believes he fell overboard—a dead body was found in the harbour today.

"Reported the attack to captain of police today, and have a great mind to publish it in the newspapers, in the hopes that the British public may know how inefficient our rule is, that such flagrant outrages as this and frequent piracies are possible here. A collection of similar facts which have occurred in the past two years in this harbour and neighbourhood, not a day's sail from our squadron of gunboats and ships of war would be startling and lead to prompt measures."

The fact is that at the time the whole long-shore population of the China Coast consisted of nothing but vast rookeries of pirates. The mysterious disappearance of the *Caliph* in the China Seas was always put down to pirates as no trace of her was ever found, and there was no record of any bad weather at the time.

The Looting of the "Young Lochinvar."

All Chinese fishermen were ready for a bit of pirating if the chance came their way. In 1866 when the *Young Lochinvar* got ashore in a fog at the entrance to the Min River, it was early morning and the fishing fleet was just putting to sea. The stranded clipper proved too great a temptation. Swarming on board, they drove her crew over the side, and soon stripped her of everything moveable. She was in ballast, coming into load new teas, and in her hold were some long pigs of lead. These were too heavy to handle, so the looters proceeded to cut them into short lengths. They even hauled the sails out of the sail locker and cut them into lengths so that they could be easily distributed. And as the clipper's crew pulled away up the river, their last view of their beautiful ship showed her masts and yards covered with "long-tails" as these wreckers proceeded to send down her sails and running gear.

In a very short time the *Young Lochinvar* was clean gutted, and the fishing sampans, loaded with spoil, pulling back to their villages.

But if the Chinese pirates did occasionally capture a British ship, it must be confessed that they were only getting a bit of their own back. Perhaps a short while before these same pirates were coolies discharging rice at Hong Kong, where it was the custom to send the ship's boys into the hold with long bamboos, with instructions to encourage the coolies to work harder by giving them a crack every now and then as they passed along.

Indeed the Britisher of the sixties had a fine contempt for the yellow races. There was no false sentiment about him. He considered himself boss wherever he went and he let other people know it.

Cutting out Ballast Lighters at Yokohama in 1867.

It was in the newly-opened country of Japan that he sometimes reckoned without his host. The Japanese are a war-like, high-spirited race, and though by the sixties the Europeans had got a foothold in the country, it was still a precarious one.

As many of the Japanese had never seen a European at that date, even in the seaports, a whole street would turn out with much laughter and gesticulation to watch one passing.

Then, if a Japanese brave happened to be about and inflamed by wine or patriotism, the unfortunate European would almost likely be attacked. "Slashing," as this mode of attack was called, caused many a man the loss of a limb or even death in Nagasaki. Yet the Europeans, especially Britishers, persisted in treating the Japs with that same contempt for an inferior race with which they treated the Chinese.

The *Norman Court* in 1872 happened to be loading Government rice in Hiogo Bay, and had a great deal of trouble with her Japanese coolies, who cared nothing for the authority of their Chinese stevedore. At last the worried mate lost his temper and took his boot to one of them. In a moment they turned upon him with their cargo hooks. Only the carpenter was by to support him, and both men were unarmed.

But the Britisher resolutely faced them and by sheer strength of will held them back, then watching his opportunity retreated up the hatchway. As the carpenter remarked afterwards: "It was nearly a case with you, my son." Indeed but for his resolute bearing, they would certainly have killed him.

The following incident will also show the careless way in which the Japs were treated. In 1867 there were several clippers lying off Yokohama, all anxious to get discharged and away to Foochow for the new teas. But shingle ballast, suitable for a tea cargo, was scarce and had to be brought in small boats from somewhere up the gulf at Yeddo.

About a dozen vessels were waiting, including *Taeping*, *Ariel*, *Taitsing*, *Chusan* and *Black Prince*, and the American ballast man was at his wit's end to please all the skippers. At last he gave out that those who could catch the lighters coming down could take them alongside their own vessels.

This resulted in a fine bit of sport especially for the boys. The gigs of the ships were manned and sent away at daylight with strong crews and spare hands to act as prize-masters. These latter were armed with pistols or revolvers, and perhaps a cutlass or musket. When the ballast fleet was sighted coming round the point, each boat dashed forward for one of the lighters. Then, as she ran alongside, the prize-master had to jump aboard, overawe the Japs and compel them to take the lighter to his particular ship.

This required a flourish of the pistol or a prick with the cutlass; but even so these cutting-out expeditions were not always a success. Occasionally the Japs were obstinate, and sailed off with the protesting prize-master, who, however, did not always protest very much, as it generally meant a run ashore and a few hours poking his nose into the social customs of the newly-opened country. But he was not employed again in the cutting-out expeditions.

Sometimes a prize master jumped short and had to be fished out of the sea, which caused much bad language, as the delay probably let in another boat. Sometimes he jumped full on top of a Jap reclining on the deck of the lighter. Then again there was trouble.

But in this happy-go-lucky way the tea clippers were eventually ballasted and hurried off to load the teas of the season.

To Japan against the N.E. Monsoon.

Of all the intermediate passages, taking rice up to Yokohama from Hong Kong was by far the most trying and unpleasant. The clippers had the full strength of the N.E. monsoon against them all the way, and it was a steady thrash to windward with hardly a favouring tack. Day after day the same vicious N.E. gale strained their hulls, split their sails, and wore their gear. Nevertheless these gallant little ships made the trip in under 20 days, though they returned in less than half that time with the monsoon behind them.

The Tea Ports.

In the old days the tea was only loaded at Canton, Whampoa, and Macao. Then Shanghai became a favourite port, if a late loading one. But when Foochow was opened it outdid the others in popularity, owing to the early date at which its teas were ready for shipment. Later in the seventies Hankow began to attract attention, but by that time the racing was practically over.

Allowances to be made in Calculating the Racing Records.

In calculating the records of the racing the time of year the ships started and the latitude of the different ports they sailed from must be taken into consideration. Thus the splendid times of the *Chrysolite*, *Cairngorm*, *Lord of the Isles*, and some of the American clippers in the fifties must be somewhat discounted by the fact that they started late in the year and had the favourable N.E. monsoon to carry them down the China Seas. Whereas in the sixties, the best times were made against the S.W. monsoon. In this respect also a big allowance must be made in favour of vessels sailing from Shanghai at the end of June or beginning of July, when they had the full strength of the S.W. monsoon against them. Their handicap, when compared with the Foochow ships, was at least five or six days, as in the S.W. monsoon Shanghai is 500 miles dead to leeward of Foochow.

The first starters from Foochow had still another advantage over their Shanghai rivals in that they got away before the S.W. monsoon was blowing its full strength; whereas the tea was never ready in Shanghai until the monsoon was at its worst.

The *Cutty Sark* invariably loaded at Shanghai and sailed at the worst season of the year, and because she could not do the trip in under 100 days, like her more favourably circumstanced Foochow rivals, she was considered by many to be either unlucky or a failure, yet she afterwards proved herself to be one of the fastest ships ever built, the only ship, in fact, which could rival *Thermopylae's* performances in the Australian trade.

The Tea Chests.

There is one other point which I must not overlook in comparing the different tea ports, and that is the size of the tea chests. In Whampoa the tea was packed mostly in 10 catty boxes or quarter chests; at Foochow one got half chests and full chests in equal proportions; whilst at Shanghai the chests were nearly all full chests.

The tea was always measured in London on arrival, and, owing to quarter chests making better stowage, there would be some £200 more freight on a vessel loaded in Whampoa and the same vessel loaded in Shanghai.

Preparations for the Race Home from Foochow.

But in the heyday of the racing, Foochow was the loading port par excellence, and the Pagoda Anchorage, just before the tea came down the river, showed perhaps the most beautiful fleet of ships the world has ever seen.

The crack ships, which were always the first to load, began to assemble about the end of April; and until the tea came down were all engaged in painting, varnishing, and smartening themselves up and in other ways, such as sheathing over their channels, preparing for the fray. Then what a sight they made when all was spick and span, with their glistening black hulls, snow-white decks, golden gingerbread work and carving at bow and stern, newly-varnished teak deck fittings, glittering brass, and burnished copper! Every ship, of course, had her distinctive mast and bulwark colours.

Ariel's masts and spars were painted flesh colour. The panels of her bulwarks and midshiphouse were pure white, with a narrow green edging and a touch of delicate pink stencilling in the centres.

Sir Lancelot's colour was pale green, Brocklebank colour, for Captain Robinson had been brought up in their service.

Some of the ships went so far as to paint elaborate landscapes or posies of flowers on their bulwark panels. But none of them could excel the Aberdeen White Star clippers, such as *Thermopylae* and *Jerusalem*, when it came to looks. Their green sides, white figureheads, white blocks, white lower masts, bowsprit and yardarms, gold stripe, and gold scroll work were the admiration of sailors wherever they went.

The amount of brass work on these tea clippers would have put a modern steam yacht to shame.

Ariel, for instance, had brass let in flush to all her bulwark rails and stanchions, inside and out. Indeed she must have shone in a blaze of fire, for, when she was in port, it took four men from 6 a.m. to 6 p.m. every day to keep her brass work clean and bright.

Norman Court, another beautifully finished ship, had a solid brass rail all round her bulwarks.

And when we remember their brass cannons, binnacles, bucket and harness cask hoops, capstan caps, ship's bells, etc., we may well imagine how they would have pleased that well-known admiral, who always wanted his ship to be "a shining, sparkling mass of burnished gold."

When all the ships had been polished up, and lay with their yards crossed and sails bent, all ready for the arrival of the tea, a day was set aside for a grand regatta, in which all the boats of of the fleet took part. This was always a great occasion, a whole holiday for the crews, with liberal prizes for the best cutters, gigs, and sailing yawls; and, naturally, the rivalry between the different ships was intense.

Loading the Tea.

The tea came down the river in sampans and the loading of it at the Pagoda Anchorage was done with all the hustle of coaling a man-of-war against time. The first lighters down distributed a ground tier to each of the first ships, after which there were two or three sampans alongside each ship until she was loaded The tea was beautifully stowed, tier on tier, by Chinamen using big mallets. It was handled day and night, Sundays included; and the officers of the first ships were relieved in their tallying by those belonging to ships which had not that pride of place, otherwise there would have been no sleep for them until the tea was in. Clarke, in his *Clipper Ship Era*, gives a very good account of this scene, which I cannot do better than quote:—

"Cargo junks and lorchas were being warped alongside at all hours of the day and night; double gangs of good-natured, chattering coolies were on board each ship ready to handle and stow the matted chests of tea as they came alongside; comfortable sampans, worked by merry, bare-footed Chinese women, sailed or rowed in haste between the ships and the shore; slender six-oared gigs, with crews of stalwart Chinamen in white duck uniforms, darted about the harbour; while dignified master mariners, dressed in white linen or straw-coloured pongee silk, with pipe-clayed shoes and broad pith hats, impatiently handled the yoke lines.

"On shore the tyepans and their clerks hurried about in sedan chairs, carried on the shoulders of perspiring coolies, with quick firm step to the rhythm of their mild but energetic 'Woo ho—woo ho—woo ho!'

"The broad, cool verandah of the clubhouse was almost deserted. In the great hongs of Adamson, Bell; Gilman & Co.; Jardine, Matheson; Gibb, Livingston; and Sassoon, the gentry of Foochow toiled by candle light over manifests and bills of lading and exchange sustained far into the night by slowly-swinging punkahs, iced tea, and the fragrant Manila cheroot."

Tugboats were scarce at Foochow in those days, and their power was very different, but the competition to secure them was generally great. The *Woosung*, an American-built boat, came down from Shanghai for the season; and, besides her, there were the paddle boat *Island Queen*, which got in a mess with *Ariel* in 1866, and the *Undine*, a screw boat.

In 1866 the tea fleet at Foochow consisted of sixteen front-rank clippers, and in the two following years there were even more.

And now that I have given a general idea of the ships, their owners, captains, and crews, and the exciting trade in which they were engaged, I will describe the clippers in detail, and make an attempt to show the wonderful way in which they raced home with the tea.

The "Falcon," First of the Improved Clippers.

The *Falcon*, which was launched by Steele in 1859, has always been considered so great an improvement on the famous heelers of the fifties, both in her lines and sail plan, that she is spoken of as the pioneer ship of a new era. If the two Steeles, her designer and builder, deserved credit, great credit must also be awarded to Captain Maxton, who superintended her building and commanded her for a couple of voyages. Indeed, it was said at the time that a great many of the improvements and innovations introduced into the *Falcon* emanated from the brain of the famous skipper. She was chiefly noted for her powers of going to windward, but she was also very fast in light winds.

The new Steele clipper did not arrive in China in time to load the new teas at Foochow, but she gave a taste of her metal by making the best run home from Shanghai.

The Tea Race of 1859.

Ship	Port from	Date Left.	Captain	Where to.	Date of Arrival.	Days Out
Fiery Cross ..	Foochow	June 9	Duncan	London	Oct. 26	139
Ellen Rodger ..	,,	,, 10	Keay	,,	Oct. 24	136
Crest of the Wave	,,	,, 16	Steele	,,	Nov. 10	147
Ziba	,,	,, 19	Tomlinson	,,	Oct. 31	134
Sea Serpent ..	,,	,, 19	Whitmore	off Plym'th	Oct. 27	130
Challenger ..	Shanghai	Aug. 6	Killick	London	Nov. 21	107
Falcon	,,	,, 23	Maxton	,,	Dec . 7	106
Stornoway ..	,,	Sept. 4	Hart	,,	Dec. 30	117
Cairngorm ..	Canton	Aug. 17	Ryrie	,,	Dec. 7	112
Robin Hood ..	Hong Kong	Oct. 1	Cobb	off the Start	Jan. 11	102
Kate Carnie ..	Whampoa	,, 25	—	London	Feb. 8	106

The year 1859 is celebrated in the annals of the tea races as being the last year in which the Americans competed in the English trade.

The American clipper, *Sea Serpent*, with her famous commander double crew and well-known reputation, is supposed to have received special inducements to load the first teas for the English market. Some declared that she received £100 down in Foochow, so certain were the shippers that she would outstrip her rivals. Another rumour was that she was to get 30s. per ton extra freight if she beat the *Crest of the Wave.*

It was a bad year in the China Seas, and all the ships made very long passages to Anjer, *Sea Serpent* passing through the Straits on 5th August, six days behind the speedy little barque *Ziba*. The first of the ships to arrive in London was the new crack *Ellen Rodger*. But somehow or other the story has arisen that the *Sea Serpent* and the *Crest of the Wave* were the first ships in the Channel, reaching the Isle of Wight simultaneously, and that the captain of the *Sea Serpent*, leaving his mate and the pilot to bring the ship up to the Thames, landed and hurried up to London by train in order to steal a march on the *Crest of the Wave* and enter his ship as the first arrival, a piece of Yankee slimness which has gained credence the world over. But a glance at the times of the different vessels will at once show the incorrectness of the yarn.

As a matter of fact, Captain Whitmore, on his arrival in the Channel, did think he had a good chance of winning, and, putting into Plymouth, actually did go up to London by train and enter his ship before her arrival in the Thames. But his disappointment, when he found the *Ellen Rodger* and *Fiery Cross* already docked, may be imagined.

The best passage of the year was made by the Aberdeen clipper, *Robin Hood*, which, however, sailed with the latter contingent at a more favourable season of the year than the Foochow ships.

The Tea Race of 1860.

By the year 1860 Foochow had become the favourite tea port, and from this date until the end of the sixties, the majority of the cracks always loaded there, leaving Shanghai, Whampoa and Macao to vessels which were past their prime or did not intend to race.

Of these old timers the *Northfleet* made the best passage home, arriving at Deal on 16th November, 114 days out from Whampoa.

The chief times from Foochow were as follows:—

Ship	Captain	Date Left	Passed Anjer	Date Arrived	Days out
Ziba	Tomlinson	June 17	July 14	Oct. 11	126
Ellen Rodger ..	Keay	June 7	—	Oct. 4	119
Falcon ..	Maxton	June 10	July 10	Sept. 28	110
Chrysolite ..	Roy	June 27	Aug. 5	Oct. 30	125
Robin Hood ..	Cobb	July 19	Aug. 22	Nov. 20	124

In 1860 two very celebrated tea clippers were built, namely the second *Fiery Cross* and the *Flying Spur*.

"Fiery Cross."

The *Fiery Cross* was built to replace the old *Fiery Cross*, which had been wrecked in 1859. She was designed by Rennie and her half model is to be seen in the South Kensington Museum.

She was commanded on her maiden voyage by Dallas, who had been so successful with the first *Fiery Cross*, then Richard Robinson had her until 1866, and under these two famous skippers she proved well nigh invincible, receiving the premium for the first vessel in dock on no less than four occasions, and being only 24 hours behind in 1864 and 1866. Besides remaining in the forefront of the racing for years longer than any other vessel, she outlived all her contemporaries.

"Flying Spur."

The *Flying Spur* was built to take the place of *Cairngorm*, and was the last clipper in which Jardine Matheson & Co., had a large interest.

She also was a very fast little ship, but, owing to not being sailed as hard as the other cracks, she did not remain very long in the first flight.

Yet she showed up so well when in company with other clippers that there is little doubt that, but for the cautiousness of her veteran skipper in carrying sail, she would have made a name almost equal to that of *Fiery Cross*. Perhaps *Flying Spur's* best performance was 73 days to Sydney, then on to China, reaching Shanghai on the 120th day out from England. Her best run on this occasion was 328 miles, and her best week's work 2100 miles. She discharged a general cargo at Sydney, and loading coal in her hold and horses in her tween decks, broke the record between Australia and Shanghai.

The *Flying Spur* cost £20,000 to build, being of teak and Greenheart, copper fastened; and she carried 1000 tons of tea when fully loaded.

The largest shareholder in her was Sir Robert Jardine, whose crest, a winged spur, gave the reason for her name. *Flying Spur* was also the name of one of Sir Robert Jardine's best racehorses, and a model of the head of this house with the crest on a shield below formed the figurehead of the vessel.

Captain Ryrie, who left *Cairngorm* to take her, owned a quarter of her, and being well on in years, preferred nursing her to the strain of cracking on through the long passage home. But that he could press her on occasion is proved by the following incident:—

One morning *Flying Spur* was snoring through the N.E. trades under all sail to royal staysails, with her lower yards just touching the back-stays.

At 11.20 a.m. a sail was sighted on the horizon ahead. This proved to be the Glasgow clipper, *Lochleven Castle*, 80 days out from Rangoon to Liverpool.

At 1 p.m. the *Flying Spur* was up with her, and as the tea clipper went foaming by, the *Lochleven Castle's* main topgallant sail went to ribbons with a clap of thunder, and her mainsail split from top to bottom; at the same moment the cook of the *Flying Spur* with all his pots and pans was washed from the galley to the break of the poop. An hour and a half later the *Lochleven Castle* was out of sight astern.

The "Lord Macaulay."

Another tea ship launched in 1860 was the *Lord Macaulay*, this vessel though never raced against the cracks, was a very fast and handy vessel, and a great favourite on the coast. She had a chequered start to her career, being originally designed as a corvette for the Russian Navy, and she had been built as far as the first futtocks when the Crimean War broke out. This naturally broke the contract, but, as she was as sharp as a wedge under water, she was finished off for the tea trade. In appearance she was heavier looking and less yacht-like than the graceful Steele creations, being frigate built, and she differed from most tea clippers in having painted ports. Though heavily rigged she set nothing above royals.

Her owners, Messrs. Monro, were of the good old type, and gave their captains a free hand, their usual words when they said goodbye at the commencement of a voyage being: "Now captain, consider the ship is yours, we leave everything to your judgement."

And a big bonus always went to a captain after a successful voyage.

She was never raced, as I have said, in the first flight, but with a captain who was a veteran on the coast she proved a very profitable ship to her owners.

The Tea Race of 1861.

Ship	Captain	From Foochow	Date of Arrival	Days out
Ellen Rodger ..	Keay	June 11	London, Oct. 10	121
Robin Hood ..	Cobb	,, 11	Liverpool Oct. 14	125
Falcon	Maxton	,, 11	London Oct. 9	120
Fiery Cross ..	Dallas	,, 14	London Sept. 23	101
Flying Spur ..	Ryrie	,, 14	Falmouth, Oct. 16	124

Ziba, *Chrysolite*, *Northfleet* and *Challenger*, sailing from Shanghai, all made passages of over 120 days.

This was the last race of two of the most noted skippers in the trade—Maxton and Dallas. Dallas retired on his laurels, but Maxton left the sea for a partnership in the firm of Phillips, Shaw, & Lowther, which was henceforth known as Shaw, Lowther, & Maxton. The withdrawal of these veterans brought Keay to the helm of *Falcon* and Dick Robinson to that of *Fiery Cross*.

At the same time, Captain Rodger, perceiving that his crack *Ellen Rodger* was being outclassed by the new ships, determined to increase her sail plan by putting her main yards to the fore, and giving her new main yards 6 feet long.

This gave her two widths more canvas on the fore and three widths more canvas on the main, with the result that her sailing was very much improved in light winds. And in 1862 she made the best time coming home of the whole fleet, which included three new ships, the *Min*, *Whinfell*, and *Highflyer*; the latter an interesting vessel in that she was built at Blackwall by R. & H. Green, and commanded by that veteran, Anthony Enright, who, after leaving the *Chrysolite*, had made himself world-famous by his wonderful passages to Melbourne in the Black Baller *Lightning*.

Flying Spur should have been first ship home this year. She was the leading ship in the channel, and when she was off Brighton, the wind being very light, a tugboat ranged alongside and asked

£100 to tow the ship to dock. Captain Ryrie refused and offered less money, whereupon the master of the tug hailed to say that he would go and tow up the *Fiery Cross*, which, he declared, was only a little way astern off the Isle of Wight. Captain Ryrie thought this was only bluff on the part of the tug's skipper and let him go. Some hours later, whilst the *Flying Spur* lay helplessly becalmed, Captain Ryrie had the mortification of seeing the same tugboat

The Tea Race of 1862.

Ship	Captain	From	Date Left	To	Date Arrived	Days Out
Fiery Cross ..	Robinson	Foochow	May 28	London	Sept. 27	122
Robin Hood ..	Mann	,,	,, 29	,,	Oct. 13	137
Min	Smith	,,	,, 31	,,	Oct. 9	131
Flying Spur ..	Ryrie	,,	June 2	,,	Sept. 29	119
Falcon	Keay	Shanghai	,, 13	,,	Oct. 13	122
Ziba	Fine	,,	,, 15	,,	Nov. 12	150
Whinfell ..	Yeo	Foochow	,, 15	,,	Oct. 13	120
Ellen Rodger ..	McKinnon	,,	,, 19	,,	Oct. 13	116
High Flyer ..	Enright	Shanghai	,, 27	,,	Nov. 3	129
Challenger ..	Macey	,,	July 9	,,	Nov. 14	128
Chaa-sze ..	Shewan	Canton	Aug. 15	,,	Dec. 15	122

steam by with *Fiery Cross* at the end of her tow rope. Thus *Fiery Cross* got first into dock, gained the premium on the freight, Captain Robinson a gratuity of £300, and her officers and crew an extra month's pay.

There was an interesting incident in this year's race, which caused more than a little surprise in shipping circles—this was the beating given to some of the cracks of the China fleet by Money Wigram's grand little Blackwall frigate *Kent*.

Four of the clippers found themselves becalmed within a mile or two of each other when one degree north of the line in the Atlantic two of them, the *Robin Hood* and *Falcon*, being bound for London, and the other two, the *Colleen Rodger* and *Queensborough*, for Liverpool.

There was not sufficient wind to give steerage way and the ships lay with their heads pointing all round the compass. In the midst of this becalmed tea fleet, and close to the *Falcon* and *Robin Hood* lay the Blackwaller *Kent*, homeward bound from the Colonies with passengers.

Signals were exchanged between the five ships and the commander of the barque *Robin Hood* asked Captain Clayton of the *Kent*, which ship happened to be nearest to him, to try and keep him company as his vessel had damaged her rudder-head. Captain Clayton at once promised to do his best.

An hour later, being about 10 a.m., a light north-east trade sprang up and all the ships stood away with every rag hung out that would draw, including foretopmast stunsails and all staysails.

The *Kent*, though she was only an 11-knot ship, was a marvel in light winds and the merest zephyr gave her steerage-way. The *Falcon*, on the other hand, had the misfortune to be suffering from ragged copper and the Blackwaller actually left her ½ knot an hour in the light breeze; at the same time the *Kent* easily held her place abeam of the *Robin Hood*.

Meanwhile the *Ellen Rodger*, with her new sail plan, appeared over the horizon astern, and, rapidly overhauling the whole fleet, passed them and was soon out of sight ahead.

For two days the trade remained light and steady and all that time the *Kent* and *Robin Hood* ran beam and beam, whilst the *Falcon* with her ragged copper, continued to drop astern to Captain Keay's chagrin. All this time the trade had held well to the eastward, but at the end of 48 hours, it began to freshen and back into its true quarter of N.E.

All stunsail booms were now sent down and preparations made for a hard thrash to windward.

This was the *Falcon's* best point and she soon began to make up her leeway. The *Kent*, on the other hand, was handicapped, as she lay over on being braced sharp up, by her big channels trailing in the water, and the smooth-sided *Robin Hood* found no difficulty in running away from her.

However, Captain Clayton had had his ambition fired by the fine performance of his vessel during the previous two days, and he determined to strain every nerve in an attempt to beat the clippers home. For the rest of the passage he scarcely left the deck, pressing the little *Kent* as she had never been pressed before. But he saw no more of the tea ships, and at length, after a good run, found himself in the channel. When off the Eddystone Light, he hove-to in order to report himself and signalled for a Plymouth pilot.

No sooner was the pilot aboard than he asked him to take his report ashore without delay, at the same time giving him a printed form on which the names of the ships spoken on the passage were entered. With the usual present of rum and tobacco, the pilot bundled overboard again, whilst the *Kent*, swinging her main-yard, rushed off up channel before a fair wind with stunsails set alow and aloft.

Next morning found the Blackwaller off Dungeness, and great was the excitement on board when two clouds of canvas were sighted astern, for it had been ascertained from the pilot that none of the ships reported by the *Kent* had arrived. It only needed a glance to tell Captain Clayton to whom those leaning pyramids of canvas astern belonged. "Here come the tea clippers," he cried to his chief officer. "Signal for a steamer at once."

The *Kent* was 5 miles off the Ness when a tug appeared, to be greeted by cheers from the excited passengers and crew of the Blackwall frigate. The wind was dead aft, and the China ships only 4 or 5 miles astern. Clayton had his stunsails in in five minutes, and the tow boat hustled her line aboard as quickly as possible. And hardly was the *Kent* fast behind her tug before the clippers, which proved to be the *Falcon* and *Robin Hood*, began to take in their kites and signal for steam. Thereupon, not being content with one tug, Clayton signalled for another, and was soon being swept up channel with a tow rope over each bow, but the tea ships were close on his heels. In the end the *Kent* just managed to beat the *Robin Hood* into the East India Dock by half an hour, the *Falcon* being close behind the *Robin Hood*.

The three vessels had parted company 27 days before, and the *Kent's* feat in beating the two flyers in the run home from the line became the talk of the city, whilst young Clayton found himself famous in shipping circles. He even had the honour of being introduced by old Money Wigram to the great Duncan Dunbar, who looked him up and down with astonishment, for he was very young to be captain of a crack London passenger ship.

Clipper Ships Launched 1861-1862.

Meanwhile several new tea ships had been launched. None of them, however, were as fast as either *Fiery Cross* or *Flying Spur*. The sister-ships *Min* and *Guinevere* were perhaps the pick of the bunch, though they did not satisfy the ambitions of either their owners or their builder, Robert Steele. They were not fine enough in the buttocks, and when at full speed heaped up a wave under the lee buttock which detracted from their pace.

The Tea Race of 1863.

Ship	Captain	From	Date Left	To	Date Arrived	Days Out
Fiery Cross ..	Robinson	Foochow	May 27	London	Sept. 8	104
Falcon ..	Keay	,,	,, 27	,,	Oct. 5	130
Min	Smith	,,	,, 28	,,	,, 5	129
Flying Spur ..	Ryrie	,,	June 1	,,	,, 5	126
Ellen Rodger	McKinnon	,,	,, 3	,,	,, 5	124
Robin Hood ..	Mann	,,	,, 4	,,	,, 5	123
Ziba	Jones	,,	,, 5	Liverpool	Sept. 19	106
Highflyer ..	Enright	,,	,, 8	London	Oct. 20	134
Challenger ..	Macey	Hankow	,, 14	,,	,, 20	128
Coulnakyle ..	Morrison	Shanghai	,, 20	,,	,, 29	131
Silver Eagle ..	—	,,	,, 22	,,	,, 30	130
Guinevere ..	M'Lean	,,	,, 27	,,	,, 28	123
Chrysolite ..	Varian	Hong Kong	July 15	,,	Nov. 14	122
White Adder ..	Bowers	Shanghai	,, 17	,,	,, 7	113
Friar Tuck ..	—	,,	,, 23	Scilly	,, 27	127

Nevertheless *Min* was the first ship home from Shanghai in 1866 and *Guinevere* distinguished herself by beating *Eliza Shaw* from Shanghai in 1864 in a dual race for large stakes. But these two ships undoubtedly taught the Steeles a great deal, for they were followed by the Greenock firm's masterpieces *Serica*, *Taeping*, *Ariel* and *Sir Lancelot*.

Green's *Highflyer* was not a success as a tea clipper. In fact she differed very little from the Blackwall frigates of the time. Her stern was heavy, her foremast too much in the bows, and owing to her want of length her main and mizen masts were much too close together.

The Blackwall firm had another try with *Childers*, but here they were again unlucky as she was wrecked in the Min River on her second voyage.

"Serica" and "Taeping."

In 1863 Messrs. Steele launched two of the most celebrated ships in the tea trade, the *Serica* on the 4th August and *Taeping* on the 24th December. Registering 59 tons more than *Serica*, *Taeping* was 2 feet shorter but 3 feet deeper, and, in point of speed she was perhaps a trifle faster all round. They were both fine handy sea-boats, very fast on light airs, and, as usual with Steele's creations, very sightly ships. Owned by such keen racing men as Findlay and Rodger, and well skippered, they were both raced for all they were worth.

"Black Prince" and "Belted Will."

Of the other clippers launched in 1863, *Black Prince* and *Belted Will* were undoubtedly the fastest, but they were neither of them ever driven, *Black Prince*, especially, being handicapped by a very careful captain, who should never have had command of a tea clipper.

Belted Will made a very fine run on her maiden passage, but she never did anything else that calls for notice.

Composite Construction.

It will be noticed that 1863 was the first year of the composite clippers. Before this date, several ships, including *Min*, *Guinevere*, *Highflyer*, and *White Adder* had been built with iron beams, but *Taeping*, *Black Prince*, *Eliza Shaw*, and *Pakwan* were the first ships engaged in the tea trade to be composite built as it is now understood.

The inventor of this method of construction is generally supposed to have been a certain John Jordan whose first effort was a schooner called the *Excelsior*, launched as far back as 1850.

Bilbe & Perry of Rotherhithe were amongst the earliest

supporters of the principle, their first composite ship being *Red Riding Hood* of 720 tons, built in 1857.

The composite construction in merchant ships was the transition stage between wood and iron. It only had a short popularity of perhaps a dozen years, and that only amongst clipper ships such as the China clippers and small Australian wool clippers. It was specially suitable, though, for the tea trade, where great strength was wanted, and in which iron ships were never popular for two reasons, firstly, that iron was considered bad for the tea, and secondly, that they could never equal wooden ships in light winds.

The Tea Race of 1864.

Ship.	Captain.	From	Date of Departure	Date of Arrival	Days Out.
Fiery Cross ..	Robinson	Foochow	May 29	London, Sept. 20	114
Flying Spur ..	Gunn	,,	June 1	,, Oct. 13	134
Serica	Innes	,,	,, 2	,, Sept. 19	109
Belted Will ..	Graham	Hong Kong	,, 3	,, ,, 20	109
Young Lochinvar	Glass	,,	,, 4	Oct. 8	126
Robin Hood ..	Darling	,,	,, 6	run down & sunk	—
Childers ..	Enright	Shanghai	,, 9	London, Oct. 21	134
Scawfell ..	Thompson	Canton	,, 7	,, ,, 13	128
Min	Smith	Foochow	,, 11	,, ,, 14	125
Ziba	Jones	,,	,, 14	Liverpool, ,, 13	121
Red Riding Hood	—	,,	July 14	London Nov. 14	123
Eliza Shaw ..	—	Shanghai	June 14	,, Oct. 21	129
Guinevere ..	M'Lean	,,	,, 17	,, ,, 20	125
Challenger ..	Macey	Hankow	,, 17	,, ,, 25	130
Yangtze ..	—	Foochow	,, 17	,, ,, 22	127
Ellen Rodger ..	M'Kinnon	,,	,, 19	,, ,, 21	124
Falcon	Keay	Hankow	,, 20	,, ,, 14	116
Kelso	—	Hong Kong	,, 25	,, ,, 24	121
Taeping ..	M'Kinnon	Shanghai	July 1	put into Amoy disabled July 23	—
Coulnakyle ..	—	,,	,, 1	put into Hong Kong disabled July 20	—
Whinfell ..	—	Foochow	,, 1	London Nov. 15	137
Silver Eagle ..	—	Shanghai	,, 8	,, ,, 11	126
White Adder ..	Bell	,,	,, 8	,, ,, 16	131
Highflyer ..	Smith	,,	,, 11	,, ,, 16	128

The composite clippers proved themselves exceedingly strong and fully able to stand the strain of hard driving without being twisted out of shape, as was the case with the American soft-wood clippers; at the same time the age to which many of them lived is little short of amazing. The system, though it had a short life amongst merchant ships, still survives in the construction of yachts.

Of the two new cracks, *Serica* won her spurs by beating the redoubtable *Fiery Cross* and taking the premium for first ship in dock. *Taeping*, however, was unfortunate. Being only launched at the end of 1863, she did not get out to China in time to load the new teas from Foochow, so she went up to Shanghai. On 1st July she left Shanghai in company with *Coulnakyle*, and beating down the China Sea both ships came to grief, being so disabled that they were compelled to put into port. *Taeping* went into Amoy, and *Coulnakyle* into Hong Kong. After repairs had been effected, *Taeping* left Amoy in October and made such good use of the favourable monsoon that she arrived in London early in January, 1865, only 88 days out.

The Tea Race of 1865.

Ship	Captain	From	Date Left	Date Arrived	Days Out.
Yangtze ..	—	Foochow	May 26	Off Plymouth Oct. 5	132
Ziba	Jones	,,	,, 27	London, ,, 7	133
Serica	Innes	,,	,, 28	,, Sept. 11	106
Fiery Cross ..	Robinson	,,	,, 28	,, ,, 11	106
Flying Spur ..	—	,,	,, 30	Off Scilly, Oct. 4	127
Belted Will ..	Graham	Macao	June 5	off Plymouth, ,, 4	121
Black Prince ..	Inglis	,,	,, 5	off Falmouth, ,, 5	122
Min	Smith	,,	,, 8	London, ,, 8	122
Young Lochinvar	Glass	Foochow	,, 9	,, ,, 7	120
Eliza Shaw ..	—	Shanghai	,, 11	off Plymouth, ,, 4	115
Pakwan ..	—	Macao	,, 14	off Lizard, ,, 4	112
Taeping.. ..	M'Kinnon	Foochow	,, 29	in the Downs, ,, 9	102

Taeping, which again made the best time of the year, was unfortunate in not being able to sail earlier, thus having no chance of gaining the premium. Her late launch in 1863 and her disablement in 1864 had thus kept her out of the first flight for two years.

Serica and *Fiery Cross*, however, ran a marvellous race. Captains Robinson and Innes arranged to be towed to sea by the same tug, and on being cast off at the mouth of the Min River made sail at one and the same moment.

After being constantly in company down the China Sea, their times at Anjer were—

Serica	23rd June, 27 days out.
Fiery Cross	24th June, 28 days out.

They were again many times in company during the rest of the passage, and finally made their numbers off St. Catherines simultaneously.

Here luck, which so often affected the end of these ocean races, came in. It was a Sunday, and a light westerly wind was blowing. Off Beachy Head the *Serica* was leading by two miles, but *Fiery Cross* had the good fortune to fall in with a tug, and being taken in tow reached Gravesend on the same tide. No other tug being about, *Serica* stood out into mid-channel in order to get the benefit of the flood, and she was up to the Foreland before a tug hove in sight and gave her a towrope, thus she just missed the tide which carried *Fiery Cross* into the river and gave her the race.

Large sums were wagered on *Serica* and *Fiery Cross*, and the result of the race caused some dissatisfaction amongst the backers, many of whom contended that the first ship to make her number should be considered the first ship home.

Of the Macao ships, *Pakwan*, to the great surprise of her captain made the best passage; but here again a very close race was sailed between *Belted Will* and *Black Prince*.

To show the inferiority of steamers at this date, the *Annette* left Macao on 31st May, and did not reach the Downs until 13th October, 135 days out, thus making a worse passage than the slowest of the clippers.

The "Ariel."

In 1865 Robert Steele built two of the most beautiful little ships that ever left the ways. These were the *Ariel* for Shaw, Maxton & Co., and the *Sir Lancelot* for MacCunn.

The *Ariel* was launched on the 29th June. Her dimensions by builder's measurement were—

Length of keel and fore-rake	195 feet
Breadth of beam	33·9 feet
Depth of hold	21 feet
Tonnage	1058·73 tons

She was launched with 100 tons of fixed iron ballast moulded into the limbers between the ceiling and the outer skin, laid along the keelson and tapering towards the bow and stern. In addition to this, 20 tons of moveable pig iron ballast was also on board. This gave her a mean draft of 10 feet.

When loaded with tea she also required about 200 tons of washed shingle. So sensitive were these yacht-like tea clippers that their proper ballasting was one of the most important considerations with their captains, and had not a little to do with their success.

During the three voyages that Captain Keay had *Ariel*, he gradually lessened the tea ballast and trimmed her more by the stern.

On her first voyage she had	340	tons with mean draft of	13 ft.	6	ins.
On her second ,, ,,	324	,, ,, ,,	13 ft.	4½	ins.
On her third .. ,,	310	,, ,, ,,	13 ft.	3	ins.

Her best trim was as much as 5 inches by the stern.

No one could be more qualified to give an account of *Ariel* than Captain Keay, who commanded her through her victorious career. I therefore have no hesitation in quoting his letter to me, in which he speaks of her with the true love of a sailor for his ship—

"*Ariel* was a perfect beauty to every nautical man who saw her; in symmetrical grace and proportion of hull, spars, sails, rigging and finish, she satisfied the eye and put all in love with her without exception. The curve of stem, figurehead and entrance, the easy sheer and graceful lines of the hull seemed grown and finished as life takes shape and beauty; the proportion and stand of her masts and yards were all perfect. On deck there was the same complete good taste; roomy flush decks with pure white bulwark panels, delicately bordered with green and minutely touched in the centre with azure and vermilion. She had no topgallant bulwarks (her main rail was only 3 feet high) but stanchions of polished teak, protected by brass tubing let in flush.

"It was a pleasure to coach her. Very light airs gave her headway and I could trust her like a thing alive in all evolutions; in fact, she could do anything short of speaking.

"*Ariel* often went 11 and 12 knots sharp on a bowline, and

in fair winds 14, 15 and 16 knots for hours together. The best day's work in south latitude, running east, was 340 nautical miles by observation, and that was done by carrying all plain sail except mizen royal, the wind being three or four points on the quarter.

"We could tack or wear with the watch, but never hesitated to call all hands, night or day, tacking, reefing, etc., in strong winds."

With regard to sail plan, *Ariel* was not overhatted, yet carried a sufficient cloud of canvas to make her a ticklish boat to handle when being heavily pressed, and Captain Keay states that she always required careful watching.

Her actual sail area was about the same as *Sir Lancelot's*. She had very long lower masts and her courses were very deep, her main tack coming right down to the deck. These big courses gave her a good pull in light and moderate winds. Like all sharp ships, she was very wet in bad weather; indeed, her petty officers could not show themselves outside the midship-house without getting drenched in anything of a blow, and they were little better off than water rats when running the easting down.

Her chief spar measurements were—

Mainmast—deck to lower masthead	62 feet
Mainmast—deck to truck	138 ,,
Mainyard—boom iron to boom iron	70 ,,
Spanker boom	50 ,,
Knightheads to flying jibboom end	67 ,,

The "Sir Lancelot."

Sir Lancelot was launched on 27th July, 1865, just a month after *Ariel*. Mr. MacCunn's first wish had been to build her 10 feet longer, but this idea was given up and her builder's measurements worked out the same as those of *Ariel*, though her net register made her 34 tons larger and her registered dimensions also differed slightly.

Like *Ariel* she was of composite construction with teak planking to bilge and elm bottom. Her ballasting also was exactly the same as *Ariel's*. She delivered 1430 tons of tea on a mean draft of 18 feet 8 inches.

No expense, of course, was spared in her outfit. In those days a crack clipper or packet was as smartly kept up as a modern yacht, and as artistic beauty was thought more of than it is now, such vessels as *Ariel* and *Sir Lancelot* were perhaps the most beautiful fabrics ever created to please the nautical eye of man. A knight in mail armour with plumed helmet, his vizor open and his right hand in the act of drawing his sword, formed *Sir Lancelot's* figure-head.

With regard to the name, Mr. MacCunn had an interesting correspondence with Tennyson, as to whether it should be spelt "Launcelot" or "Lancelot". Tennyson wrote that it should be "Lancelot." I mention this as I notice that the name is so often spelt wrongly, an instance of which is to be found in Captain Clarke's *Clipper Ship Era.*

Another mistake which I should like to correct is the 45,000 square feet of canvas, credited to *Sir Lancelot* by Lindsay in his *Merchant Shipping*, which has been followed by several other writers.

Mr. James MacCunn has kindly supplied me with *Sir Lancelot's* sail plan in detail, which will be found in the appendix. This shows that *Sir Lancelot's* net sail spread worked out at 32,811 square feet, not over 40,000 as has been stated. This table, however, takes no account of such flying kites as the Jamie Green, ringtail, water-sails, bonnets and wings to lower stunsails, which were always made by the sailmaker on board according to the directions given him by the captain. When all these auxiliaries were set there was an addition of from 2000 to 2500 square feet of canvas to the full working suit of the ships.

"Sir Lancelot's" Unfortunate Maiden Voyage.

Things rarely go absolutely right on a vessel's maiden voyage, and thus it is very necessary to place a new ship under the command of an experienced captain. In this respect *Ariel* was lucky in having Captain John Keay who, after putting the various small mistakes of riggers and builders right, learnt to know his ship thoroughly and had her tuned up to the nines by the time she reached Foochow to load the new teas.

But that the voyage until then had not been without the usual maiden passage incidents, *Ariel's* abstract logs showed only too clearly.

It was not, however, until her second voyage that *Sir Lancelot* had a chance to show what she could do.

Mr. MacCunn was unfortunate in being unable to obtain a good skipper for her at the start, and whilst *Ariel* under Captain Keay made her reputation on her first voyage, *Sir Lancelot* was handicapped out of the running by a bad jockey. Mr. MacCunn lost the skipper whom he had in his mind owing to a slight delay in the completion of *Sir Lancelot*, the man being snapped up by another firm. So a captain named Macdougall with good credentials was given command.

This man showed soon that he was totally unfit to command such a thoroughbred of the seas as a China clipper, and, possessed of neither nerve nor go, he blundered through the entire voyage.

After a protracted intermediate passage to Bangkok and back to Hong Kong with rice, the *Sir Lancelot* reached Hankow and was lucky enough to be loaded by Jardine, Matheson & Co., at £7 per ton. But at Hankow a collision, which damaged her main rigging and head gear and lost her anchors and cables, destroyed the last remnants of her captain's nerve.

Instead of bravely facing the intricate navigation of the China Seas and taking the Anjer route home according to Mr. MacCunn's instructions, Macdougall chose the Eastern Passage *via* Ombai. After wandering aimlessly along in continual calms, he took 42 days to clear Sandalwood Island. And even in this drifting performance he managed to show his incompetence by carrying away the slings of his mainyard. Yet no log book ever contained so many references to shortening sail, "in topgallant sails," etc., appearing whenever there was a good breeze and a chance to get ahead.

Captain M'Lean of the *Guinevere*, which had been wrecked in the Yangtse, had a passage home on the *Sir Lancelot*, and whether these two incompetents liquored up too much or not, they managed to make a complete hash of the passage, with the result that the gallant clipper took 122 days to struggle home.

It was thus that the *Sir Lancelot* missed the great tea race of 1866, in which her sister ship *Ariel* performed so brilliantly.

The Great Tea Race of 1866.

It is probable that no race ever sailed on blue water created so much excitement as the great tea race of 1866. For some years past the public interest had been growing, until it had now come to pass that even those who dwelt in sleepy inland villages looked eagerly down the shipping columns of the morning papers for news of the racing clippers. And if this interest was shown by landsmen who had no connection with either the ships or the trade, it is not surprising that the great shipping community of Great Britain looked upon the tea races much as the British public look upon the Derby or the Boat Race.

Every man with a nautical cut to his jib had a bet upon the result, whilst the rival owners, agents and shippers wagered huge sums. Nor were the captains and crews of the vessels themselves backward in this respect. I have already related how the crews of those old antagonists *Fiery Cross* and *Serica* wagered a month's pay against each other. As for the captains, it had come to be almost a form of etiquette on the China Coast for a captain to back his own ship. I shall not forget the scorn in the voice of one of these old tea clipper captains, when, in describing a race to me, he remarked that he could not get his opponent to wager even the customary beaver hat. One other captain I know of who steadily refuse to bet and that was the famous Anthony Enright, when in command of *Chrysolite*. He refused from religious scruples.

In 1866 freights were up to £7 a ton for the first six or eight ships loading at Foochow, and the beginning of May found sixteen of the best known and finest clippers assembled at the Pagoda Anchorage, waiting for the first season's tea to come down the river.

Amongst these were the *Ariel*, *Fiery Cross*, *Serica*, *Taeping*, *Falcon*, *Flying Spur*, *Black Prince*, *Chinaman*, *Ada*, *Coulnakyle*, and *Taitsing*, Findlay's new crack, whose chances were fancied by many a shrewd judge.

The beautiful *Ariel* lay below the rest of the fleet, close to the Pagoda Rock, Though no betting prices were published and no brazen-tongued bookies were present to cry the odds, she was recognised as the favourite, and, as such, favoured by the shippers.

As a rule in the tea races, the first ship to finish loading was that which the shippers believed to be the fastest.

Occasionally they were mistaken, as in the case of *Maitland*, which they made favourite in the 1867 race; occasionally also, one of the cracks arrived too late to get away amongst the first ships.

On 24th May the first lighters of new tea came down the river and *Ariel* stowed her flooring chop of 391 chests and 220 half-chests. On Sunday, the 27th May she had sixteen lighters alongside, and Chinese coolies worked day and night getting it aboard, until at 2 p.m. on Monday, 28th May, the last chest was landed, 1,230,900 lbs. of tea being on board.

The cargoes of the other fancied ships were—

Fiery Cross	854,236	lbs.
Taeping	1,108,700	„
Serica	954,236	„
Taitsing	1,093,130	„

Ariel was the first ship ready. At 5 p.m. on the 28th she unmoored, and, with the tug-boat *Island Queen* alongside, dropped well below the shipping and anchored for the night. The next ship to unmoor was the *Fiery Cross*, twelve hours later, and she in turn was followed by the *Taeping* and *Serica*, which finished loading and got away together, then came *Taitsing*, a day behind. But, as was proved in this instance, the first vessel to finish loading was not always the first vessel to clear the river.

At 5 a.m. on Tuesday, 29th May, the *Ariel* hove up and proceeded down the river with the tug towing alongside. At 8.30 when nearing Sharp Rock, she discharged her Chinese river pilot and the tug was sent ahead to tow.

The scenery of the Min River is magnificent; on each side hill rises above hill, here cultivated to the summits by means of terraces, there so steep as to admit only of stunted fir. Along the banks quaint Chinese villages, forts and joss houses are scattered, whilst on the surface of the water wood-laden junks and all kinds of river craft are plentiful. But like most beautiful rivers, the Min is a swift-running one, the tide becoming a regular sluice wherever the channel narrows, and in those days tug boats on the China Coast were far from being what they are now. Thus it came about that no sooner was the *Island Queen* ahead than she began to sheer about so wildly in the "chow-chow" water as to be quite unmanageable and the *Ariel* was compelled to let go her anchor in a hurry in order to prevent a disaster. Once more an attempt was made with the tug lashed alongside. Going outside the wreck of the *Childers*, every ounce of steam was put on in a vain attempt to save the tide. But again the tug made a mess of it and there was enough top on the water to cause considerable damage not only to the *Island Queen's* sponsons, but to the *Ariel's* glossy black side through the grinding of the two vessels together. Captain Keay wanted to try the tug ahead again, but it was now dead low tide and the pilot refused to go on until there was more water, so there was nothing to do but anchor.

Captain Keay's mortification at this unfortunate start was further increased by the sight of *Fiery Cross* coming down with a good tug ahead. The *Ariel* was drawing 18 ft. 5½ ins. on a mean draft, but the *Fiery Cross* drew considerably less and thus was able to proceed to sea without delay and went by with three mocking cheers of farewell.

That night *Ariel* was again delayed by the fates, for though there was enough water for her to proceed by 8.30 p.m., the weather was so thick and showery that the pilot refused to venture. However, there was a full night's work aboard, as the *Ariel* was badly out of trim, being no less than 5 inches by the head; this no doubt, had not made the tug's difficulties any easier. Everything of weight including cables, hawsers, salt provisions and fore-peak stores, was cleared out from forward and brought aft of the mizen mast, in addition to which the captain's cabin was filled with tea. This, however, did not entail much hardship on the *Ariel's* commander, for he was rarely off the deck.

At 9 a.m. on the 30th *Ariel* once more got under weigh, the tug going ahead with a hawser from each quarter. But the delay had given *Fiery Cross* a lead of 14 hours, and also brought up the *Taeping* and *Serica*, which were only a few minutes behind *Ariel* in crossing the bar. By 10.30 a.m. all three ships, *Ariel*, *Taeping* and *Serica* were outside the Outer Knoll and hove-to in order to drop their pilots. Once more the luck was against *Ariel.* The wretched *Island Queen,* in lowering her boat to fetch off the pilot, capsized her and was so long in saving her boat's crew, who were struggling in the water, that the *Ariel* had at last to signal for a pilot boat to come and take off her pilot.

At last, at 11.30 a.m., Captain Keay filled his mainyard and stood away S. by E. ½ E. for Turnabout Island. There was a moderate N.E. wind. All three ships set main skysails and fore topmast and lower stunsails. It was as level a start as could be wished for three favourites. For a time the vessels kept close to each other, the *Ariel* slowly gaining on both *Taeping* and *Serica.* But the weather was thick and rainy and before nightfall the three racers had lost sight of each other.

It is now time to mention the other competitors; *Taitsing,* the next away, left the river at midnight on the 31st May, then came *Black Prince* on the 3rd June, followed by *Chinaman* and *Flying Spur* on the 5th, and *Ada* on the 6th, and at still later dates came the rest of the fleet including *Falcon, Coulnakyle, Yangtze, Belted Will, Pakwan, White Adder* and *Golden Spur.* The race to be first home, however, was confined to the first five starters, as none of the later ships had a chance of outstripping them.

Captain Robinson with his six year-old veteran, *Fiery Cross,* led the way to Anjer, following the usual route down the China Seas, through the Formosa Channel to the Paracels, after which, with the fickle S.W. monsoon of June, it was a case of courting the land and sea breezes down the Cochin China Coast, then crossing to the Borneo Coast and repeating the operation. Knowing captains showed great skill in tacking in and out, so timing it that they were well under the land about the hour that the land breeze was due to spring up, which was often not until the middle of the night. The ships usually went through the Api Passage and then south to Anjer by the Gaspar Strait.

This first part of the race was the most trying part. Every advantage had to be taken of the faintest ripple on the water. In the S.W. monsoon there was no settled weather, the wind now came with a rush, now died away to nothing, and it would chop about in squalls so quickly that it was almost an impossibility to avoid getting caught aback, and woe betide the vessel caught aback with flying kites aloft. This was the most fruitful source of losing spars and sails in all the varied weather of the long passage.

It would be baking hot with a scorching calm and glaring sun one moment, and the next a squall would sweep up out of nowhere, accompanied by a cascade of rain—the wind would come with the strength of half a gale in one squall and in the next perhaps there would be no wind at all but only a blinding sluice of rain. The navigation too was tricky and strewn with faultily charted reefs; rocks blocked the fairway of narrow channels, whilst the currents generally behaved contrary to the sailing instructions or appeared where none had been before.

Under such conditions it is not surprising that the captains rarely left their quarter-decks for more than a few minutes at a time until Anjer was passed.

Fiery Cross had fair N.E. winds to 19° 20′ N.; but here a few hour's calm and southerly airs allowed *Ariel* and *Taeping* to close up on her. Then the N.E. wind came away again and carried her safely past the Paracels. *Ariel* had N.E. winds to 17° 51′ N., then in her turn had to fight with light airs and calms.

She passed the Paracels on the same day as *Fiery Cross*, 3rd June, the North Shoal bearing south 8 miles at 1 p.m. *Taeping* also passed them on that day, but *Serica* had dropped back and was about a day behind.

Though the three leaders were so close together on the 3rd, they did not sight each other, but *Taeping* and *Ariel* had been in company on the previous day. After the Paracels they had to contend with the bothersome S.W. monsoon.

On 8th June *Taeping* and *Fiery Cross* passed each other on opposite tacks, and on the following day in 7° N., 110° E., *Taeping* and *Ariel* exchanged signals, *Taeping* being 3 miles to leeward at 5 p.m. Both ships held on the same tack through the night, and when morning broke *Ariel* had the satisfaction of seeing that she had weathered on her rival a little, *Taeping* being 4 miles on her lee quarter.

After this the ships did not meet again before Anjer, which was passed in the following order:—

1. *Fiery Cross*	on June	18 at	noon	.. ..	21 days out.
2. *Ariel*	,,	20 ,,	7 a.m.	.. ..	21 ,,
3. *Taeping*	,,	20 ,,	1 p.m.	.. ..	21 ,,
4. *Serica*	,,	22 ,,	6 p.m.	.. ..	23 ,,
5. *Taitsing*	,,	26 ,,	10 p.m.	.. ..	26 ,,

It was between Anjer and Mauritius under the influence of the steady S.E. trade wind of the Indian Ocean, that the racing tea ships were accustomed to make their best times. It was on this stretch that every kind of flying kite was set and hung on to until the very last moment.

Ringtails were bent outside the spanker with a watersail underneath the boom; wings in the shape of spare mizen staysails were laced on to the lower stunsails; the Jamie Green was almost a fixture along the bowsprit; a bonnet was laced on the foresail, watersails hauled out under the passeree booms and a spare flying jib run well up on the fore royal or fore topgallant stay as a jib topsail. Every stay to the main skysail had its staysail, whilst the fore topmast and main topmast staysails were so voluminous that their heads came to the collars of their respective stays. Then the large suit of stunsails, from the royal down, were sent aloft, tea clippers having as a rule a large and small suit.

The following entry on 26th June in *Ariel's* log gives some idea of the carrying on:—"Carpenter making stunsail yards, having carried away two topmast, one royal and one topgallant stunsail yard."

The best runs of the passage were made on this stretch and were—

			Miles.	
On June	24 *Fiery Cross*	.. ..	328	
,, ,,	26 *Ariel*	.. ..	330	
,, ,,	25 *Taeping*	.. ..	319	*Ariel* 317.
,, ,,	29 *Serica*	.. ..	291	
,, July	2 *Taitsing*	.. ..	318	

The ships passed the longitude of the Mauritius without changing their order, as follows:—

1. *Fiery Cross*	on June	29	.. ..	10 days from Anjer.
2. *Ariel*	,, July	1	.. ..	11 ,, ,,
3. *Taeping*	,, ,,	1	.. ..	11 ,, ,,
4. *Serica*	,, ,,	4	.. ..	12 ,, ,,
5. *Taitsing*	,, ,,	9	.. ..	13 ,, ,,

From Mauritius to the Cape all kinds of weather were encountered, from light airs to heavy storms.

Fiery Cross, steering into the Natal Coast to make the most of the Agulhas current, actually sighted the Cape, but *Ariel* and *Taeping*, further to the southward, were luckier with their winds, and closed up on her, whilst *Serica* steering a more southerly course than the others got caught by the westerlies and had to tack up north again in order to get the favourable current.

Whilst rounding the African Coast the tea ships had a chance of showing their paces against other fast homeward bounders. On 10th July, with a fresh southerly breeze, *Ariel* passed one of Smith's smart little Cities, the *City of Bombay*, going nearly 2 feet to her one, and on 15th July, with a light N.W. wind, she overhauled the Donald Currie flyer, *Tantallon Castle*, with the greatest of ease.

The meridian of the Cape was passed in the following order:—

1. *Fiery Cross* on July 15 at 10 p.m. 16 days from Mauritius, 47 days from Foochow.
2. *Ariel* on July 15 a few hours later, 14 days from Mauritius, 44 days from Foochow.
3. *Taeping* on July 16 half a day later, 15 days from Mauritius, 45 days from Foochow.
4. *Serica* on July 19, 15 days from Mauritius, 50 days from Foochow.
5. *Taitsing* on July 24, 15 days from Mauritius, 54 days from Foochow.

From the Cape *Fiery Cross* and *Ariel*, steering the same course, had light winds, whilst *Taeping*, some 300 miles nearer the African Coast, had better luck.

On 19th July, all three ships were abreast though out of sight of each other, and *Taeping*, continuing to gain, was the first to pass St. Helena. Meanwhile, *Serica*, following in the wake of *Taeping* was making the best time of the whole fleet, and actually went ahead of *Ariel* for a day or two.

At St. Helena the order was—

1.	*Taeping*	on July	27	11 days from the Cape.	
2.	*Fiery Cross*	,, ,,	28	13 ,,	,,
3.	*Serica*	,, ,,	29	10 ,,	,,
4.	*Ariel*	,, ,,	29	14 ,,	,,
5.	*Taitsing*	,, August	5	12 ,,	,,

However, between St. Helena and Ascension *Ariel* got a slant and making up a day, again drew level with *Fiery Cross*; at the same time *Taitsing*, which had lagged behind for so long, began to make up ground.

The timing of the fleet at Ascension was—

1.	*Taeping*	on July	31	4	days	from	St. Helena.
2.	*Fiery Cross*	,, August	1	4	,,		,,
3.	*Ariel*	,, ,,	1 at 4.30 p.m.	3	,,		,,
4.	*Serica*	,, ,,	2	4	,,		,,
5.	*Taitsing*	,, ,,	8	3	,,		,,

At the equator *Fiery Cross* and *Ariel* had again drawn level with *Taeping*, all three vessels crossing the line on the same day, the order being—

1.	*Taeping*	on August	4	4	days	from	Ascension.
2.	*Fiery Cross*	,, ,,	4	3	,,		,,
3.	*Ariel*	,, ,,	4	3	,,		,,
4.	*Serica*	,, ,,	6	4	,,		,,
5.	*Taitsing*	,, ,,	12	4	,,		,,

From the line the racing became closer than ever.

On 9th August in lat. 12° 29′ N., the *Taeping* and *Fiery Cross* exchanged signals, the *Ariel* being then just a day behind them and further to the westward. But during the next few days the latter again resumed the lead.

Taeping and *Fiery Cross*, with light and variable winds, remained in company till 17th August, their noon position on that date being 27° 53′ N., 36° 54′ W. Here bad luck fell to the share of *Fiery Cross*, for whilst she lay in a dead calm she had the mortification of seeing *Taeping* pick up a fresh breeze, which carried the latter out of sight in four or five hours, the *Fiery Cross* remaining becalmed and not making a knot an hour for 24 hours.

Meanwhile the times of passing San Antonio, Cape Verde, had been—

1.	*Ariel*	on August	12	8	days	from	equator.
2.	*Taeping*	,, ,,	13	9	,,		,,
3.	*Fiery Cross*	,, ,,	13	9	,,		,,
4.	*Serica*	,, ,,	13	7	,,		,,
5.	*Taitsing*	,, ,,	19	7	,,		,,

It will thus be seen that *Serica* had run up to *Taeping* and *Fiery Cross*, whilst *Taitsing* had also shortened her distance from the leaders by a couple of days.

As the ships neared the Western Isles, the ranks continued to close up and on 29th August the first four ships passed Flores within the 24 hours, *Ariel* still holding the lead from *Fiery Cross*, their noon positions being—

Ariel 41° 54′ N., 34° 32′ W. *Fiery Cross* 41° 5′ N., 35° 51′ W.

But the most remarkable incident in his stretch was the wonderful sailing of *Taitsing*, which had made up three days on the leading ship.

The times here are truly astonishing in their closeness—

1.	*Ariel*	on August 29	..	17 days from San Antonio	91 days out.
2.	*Fiery Cross*	„ „ 29	..	16 „ „	92 „
3.	*Taeping*	„ „ 29	..	16 „ „	91 „
4.	*Serica*	„ „ 29	..	16 „ „	91 „
5.	*Taitsing*	„ Sept. 1	..	13 „ „	93 „

From the Western Isles the racers had fresh W. and S.W. winds, with only one day of easterly weather, all of them making the run to soundings in six days.

At 1.30 a.m. on 5th September. *Ariel* sighted the Bishop Light, and, with all possible sail set, tore along for the mouth of the Channel. At daybreak a vessel was seen on her starboard quarter carrying a press of sail.

"Instinct told me that it was the *Taeping*," Captain Keay wrote in a letter to me; and so it proved to be.

All day the two ships surged up Channel together, going 14 knots with royal stunsails and all flying kites set, the wind being strong from W.S.W.

The Lizard Lights were abeam at 8 a.m. and Start Point at noon. Towards 6 p.m., when off Portland, both ships were compelled to take in their Jamie Greens in order to get the anchors over. At 7.25 p.m. St. Catherines bore north 1 mile, and soon after midnight Beachy Head was abeam, distant 5 miles.

All this time there had been no alteration to speak of in the distance between the two vessels—*Ariel* kept her lead, gaining a little as the wind freshened and letting *Taeping* up again as it took off.

At 3 a.m., when nearing Dungeness, *Ariel* began to reduce sail, send up rockets and burn blue lights. At 4 a.m., when abreast of the light and 1½ miles off, she hove to, still signalling with flares and rockets for a pilot.

At 5 a.m., *Taeping* was close astern of *Ariel* and also signalling, but as she showed no signs of heaving to, Captain Keay began to fear that she meant to run ahead of him, he therefore bore up athwart her hawse, determined at all costs to prevent M'Kinnon from nicking in and getting the first pilot. This daring manoeuvre succeeded. M'Kinnon at once gave in and hove to.

At 5.30 a.m. the pilot cutters were seen coming out of the Roads and Captain Keay at once kept away and laid *Ariel* in between *Taeping* and the cutters.

At 5.55 a.m. the pilot stepped aboard the *Ariel* and saluted Captain Keay as the first ship of the season from China.

At 6 a.m. both ships stood away for the South Foreland with their pilots aboard. *Ariel* set all plain sail but *Taeping* sent aloft topmast, topgallant and lower stunsails on one side. With this extra canvas she managed to close up a little on her rival, but *Ariel* was still a mile ahead when M'Kinnon, after shifting his stunsails across to the port side on hauling up through the Downs, was at last compelled to take them in off Deal.

Here both ships took in their white wings and signalled for steam, with their numbers flying from their peak halliards. This time it was *Taeping's* turn to crow, for the best tug coming out from behind the Ness sent her towline aboard the nearest of the two clippers, which, being the sternmost one, was of course *Taeping*. *Ariel* had to put up with a poor tug which was waiting in the Downs. Captain Keay would have taken a second tug alongside, but there was no object in the extra expense as in any case they would have to wait at Gravesend until the tide made.

Taeping, with her superior tug, arrived off Gravesend 55 minutes ahead of *Ariel*; but the latter avoided anchoring by taking another tug alongside. As soon as there was enough water both vessels proceeded. At 9 a.m. *Ariel* arrived outside the East India Dock gates. *Taeping*, having further to go, did not reach London Docks until 10 p.m., but drawing less water than *Ariel*, she was able to go through the lock and thus docked 20 minutes before the *Ariel*.

Such a close and exciting finish had never been seen before in an ocean race, and the interest it aroused caused the newspapers to vie with each other in publishing sensational accounts, and all kinds of incorrect reports as to which had won the premium, set forth as 10s. per ton in the bills of lading, were set abroad.

I therefore quote from Captain Keay's private journal in order to show conclusively how the difficulty was settled. He writes as follows:—'When the ships were telegraphed through the Downs, the owners and agents of both met and discussed the position and prospects as to who should dock first, the risk of losing the extra 10s. per ton if both should dock at the same time, or if a dispute should arise as to which was entitled to the extra freight—also that one might outwit the other by going into the Victoria Dock. It was arranged after much going and coming, that each ship should make for her respective dock and let the one which had the advantage of a few minutes claim, while the other would avoid all pretence to claiming lest the tea merchants should have power to maintain that there was no first ship as both claimed the prize—this the merchants were quite prepared to do especially as the teas were selling at a great loss."

This arrangement was adhered to; *Taeping* claimed and received the 10s. per ton, which she divided with *Ariel*; Captain M'Kinnon at the same time dividing the £100, given to the captain of the winner, with Captain Keay.

Meanwhile another of the racers was close on the heels of the dead-heaters, and whilst the tea samples were being tossed ashore from the *Ariel* and *Taeping* at midnight, *Serica* was being hauled into the West India Dock. It appeared that whilst the first two ships were racing neck and neck along the English coast, *Serica* had been tearing up the French side of the Channel, and, passing through the Downs at noon, got into the river on the same tide and just managed to scrape into the West India Dock at 11.30 p.m. as the gates were being closed.

Surely a more marvellous race could hardly be imagined. Leaving the Min River on the same tide, *Ariel*, *Taeping* and *Serica* had docked in the River Thames on the same tide. It was a proud day for Scotland, for all three captains, Keay of *Ariel*, M'Kinnon of *Taeping* and Innes of *Serica* hailed from the Land o' Cakes.

And what had become of the *Fiery Cross* which had held the lead for so long? She was only a little over 24 hours behind. At 10 a.m. on the 7th September she sighted the Isle of Wight. The wind, which had been fresh from the W.S.W., now increased to a gale, and on her arrival in the Downs it was blowing so hard that she was compelled to anchor, and was unable to get into London dock until 8 a.m. on Saturday, 8th September.

Taitsing, the last of the five, arrived in the river on Sunday, 9th September, in the forenoon.

Thus the final times were—

Ariel arrived in the Downs at	8 a.m. Sept. 6 ..	..	99	days out.
Taeping ,, ,,	8.10 a.m. Sept. 6	..	99	,,
Serica ,, ,,	noon Sept. 6 ..	..	99	,,
Fiery Cross ,, ,,	during the night Sept. 7		101	,,
Taitsing ,, ,,	forenoon Sept. 9	..	101	,,

None of the ships that sailed later approached these times.

The last arrival was the leisurely *Black Prince*, whose performance was a great disappointment to her owners and builders. As a matter of fact she was a very speedy ship, but Captain Inglis was too cautious a man ever to make a fine passage; he took her round instead of through the narrow passages such as Stolzes in Gaspar Strait, and always ran her away to leeward in a squall instead of luffing her through it.

It was on this occasion that the *Black Prince* received the name of "The Whipper-in" in the City, but Captain Inglis was impervious to chaff, and made no effort to get this title revoked, and "The Whipper-in" *Black Prince* remained for the rest of her racing days.

"Titania."

Whilst *Ariel*, *Taeping* and *Serica* were making history, Robert Steele was building the *Titania* which, in both beauty and speed, was to rival those incomparable sister ships, *Ariel* and *Sir Lancelot.*

If there was a fault to be found with these two lovely little clippers, it was that they were a little tender and in squally weather required very careful watching. When Shaw, Maxton & Co. gave

Steele the order for *Titania*, they asked him to give them a stiffer boat, and he responded by giving the new ship more beam.

Comparing the registered measurements of the three, the number of beams to length works out as follows—

Ariel	5·84
Sir Lancelot	5·83
Titania	5·55

Titania was launched on 26th November, 1866, her dimensions by builder's measurement being—

Length of keel and fore rake	199 feet
Breadth of beam	36 ,,
Depth of hold	21 ,,
Tonnage	$1222\frac{89}{94}$ tons

Her best point of sailing proved to be with the wind just abaft the beam, when it was not too strong to prevent her carrying all plain sail. She was a splendid sea boat and handled like a top; she was very lively, and, like most tea clippers, threw the water all over her in heavy weather. In light airs she cut along like a knife as long as there was a ripple on the water and it required an absolutely flat calm to stop her steering. There is no doubt that she was quite as fast as *Ariel* and *Sir Lancelot*, though it was only under Captain Burgoyne that she was allowed to show her paces.

Captain Deas, who took her from the builder's hands, should never have had a tea clipper. He was a first-class master but not a racing man, and when he took in sail it was not made again in a hurry. However, she had a good mate in Duncan, who was in the *Ariel* her first voyage and later in the *Norman Court*, and a wonderful bo's'n with only one eye, who saw as much with that eye as most men could with a dozen.

"Titania's" Disastrous Passage Out in 1866-7.

I have already described how the *Titania* was dismasted just north of the Cape Verd. The squall struck her about 8 a.m., when the man who had been sent aloft to take in the fore-topgallant stunsail was still aloft, and he had only just reached the deck when the foremast buckled just above the mast coat. As the mast went over the side it broke again where it smashed in the rail.

The *Titania's* masts like those of *Sir Lancelot* and *Ariel* were of iron, but for some reason or other the angle irons had been omitted in her case and this was given out as the reason why the foremast went. In a moment the beautiful little vessel was a wreck aloft, but luckily the hull sustained no damage. All hands were at once called to clear away the wreck and it took the carpenter three days, cutting through the buckled iron foremast. Saving what he could of the spars, Captain Deas proceeded to rig jury masts and then made the best of his way to Rio.

Here *Titania* was delayed some time whilst a wooden foremast was being built for her. When she did resume her voyage she had no sooner got down into easting weather than another defect was discovered aloft. She was about the meridian of the Cape, when the lower main masthead was found to be fractured. Sail was at once reduced and the masthead fished, the topgallant mast being sent on deck to relieve the strain upon the cap. She thus had to make the rest of the passage under easy canvas and was a long time getting out to Shanghai.

Whilst running north in the China Seas she nearly finished her own career as well as *Ariel's* through the fault of her officer of the watch.

Captain Keay was already on his way home. At 2 a.m. on the morning of 29th June, when in lat. 10° N., long. 110° E, he was beating south against the monsoon, the wind being fresh and squally, when he was nearly run down by *Titania*, which should of course have given way to him in accordance with the Rules of the Road. Captain Keay's entry in his log runs as follows—

"2 a.m.—Had to keep off for a running ship to avoid collision. Had lost his main-topgallant mast. Had double topsails and asked us, 'What ship is that?' I reproved his lubberly conduct in not hauling up to go astern of us and did not have time to answer him. Was it the *Titania*?"

It was the *Titania*, and we may guess what Captain Keay had to say to Captain Deas and his old mate Duncan when next they met.

On her arrival in Shanghai *Titania* had her main and mizen masts removed and new iron masts put in with proper angle irons.

She was, of course, too late to take any part in the racing, but getting home to London before the end of the year, was then given a complete new outfit of spars, and in January, 1868, sailed again for China with Captain Deas still in command.

"Sir Lancelot" dismasted on her Passage Out in 1866-7.

The construction of iron masts and spars was in its infancy at this date, with the result that many a vessel suffered dismasting. And *Sir Lancelot* on her second voyage was among the victims, owing to lack of structural strength in her bowsprit.

At the end of her mismanaged first voyage, Messrs. MacCunn, with many a scathing epithet, discharged their incapable skipper, and, determined to secure a first-rate man at all costs, offered Captain Robinson, who had commanded the *Fiery Cross* for two years and was in the very front rank of racing skippers, a handsome inducement to leave the old flyer and take charge of the untried *Sir Lancelot.* To Campbell's loss and MacCunn's gain, he accepted the offer, and thus it was that *Sir Lancelot,* with a new captain and picked crew, left the Thames at the beginning of December, 1866, with every anticipation of a prosperous voyage.

The start, however, was a bad one. From the first the "Flying Horse" clipper had dirty weather and was compelled to beat down Channel against a strong sou'-west blow, which resolved itself into a heavy gale as soon as she was abreast of Ushant.

On 13th December, with the wind increasing with every squall, Captain Robinson wore his ship to the south'ard and reefed down. At 3.30 p.m. both foresail and mainsail were hauled up and made fast.

At 4.30 *Sir Lancelot* was head-reaching comfortably on the starboard tack when she was struck by a tremendous squall. The gallant clipper stood up to it manfully; but, of a sudden, in the midst of the hurly-burly of screaming wind and hissing seas, there arose the sinister sound of cracking spars and tearing canvas—the bowsprit had carried away inside the forestay band close to the knight-heads. Before Robinson had time to issue a command, the foremast followed it over the side, breaking off like a carrot just above the mast coat. Then the mainmast went. This mast broke below the main deck and tore a big hole in the deck itself as it fell over the side. Next came the mizen's turn, here everything went except the lower mast, even the cross-jack yard going overboard with the rest of the wreck.

The plight of the beautiful ship may be imagined, as she rolled in the trough of the sea, her main deck gaping open and the whole fabric of her immense sailspread a tangle of broken spars, torn canvas and twisted cordage, part of it blocking up her decks and the rest pounding alongside to leeward.

All that night the *Sir Lancelot* lay helpless with only her mizen lower mast standing, whilst the mass of wreckage acted like a battering ram against her port side.

All hands worked frenziedly with axe, hatchet and saw cutting this wreckage free of the ship. And there was more than sufficient cause for haste as the *Sir Lancelot* was drifting down on the worst lee shore in the world. By daybreak Ushant bore S.E. 30 miles, but luckily for the crippled ship the wind had got further to the south'ard. All day the work of clearing the wreck and fitting jury masts went on, and so smart were both officers and crew that they had got a jury mast and jibboom rigged by the morning of the 15th, and Captain Robinson was able to get his ship away before the wind with a fore-topgallant sail, royal and staysails set forward.

At 10.30 p.m. he made S. Anthony and sailing into Falmouth without any assistance, let go his anchor in Carrick Roads. For this fine piece of seamanship the underwriters awarded £250 to be divided in proportion between Captain Robinson, his officers and crew.

But though the ship was saved, there were yet innumerable difficulties to be overcome if she was to reach China in time for the tea season. At first it was proposed to tow the lame duck to London to refit, but in the end it was decided to carry out the work at Falmouth. For this purpose Mr. James MacCunn hurried down to Cornwall to take charge whilst Captain Robinson was away looking for new masts and gear.

The resources of Falmouth were taxed to the utmost, and Mr. MacCunn found it necessary to bring down a gang of Liverpool riggers, headed by a master-rigger named Nicholas, who soon proved himself invaluable, a first rate man of go and grit whom nothing dismayed. (Many years afterwards Nicholas had command of the *Sir Lancelot* and later still became second officer of the Cunarder *Umbria.*)

There was no time to build new iron masts, but luckily Messrs. Money, Wigram & Sons were able to supply a magnificent set of Oregon pine sticks, which they sent off to Falmouth with admirable despatch.

Meanwhile, with Captain Robinson away in London, Mr. MacCunn found that he had a stiff contract before him. First the cargo had to be taken out, the undamaged part being stored in the dock warehouses, and the damaged part returned to the shippers for repairs and renewal. As soon as the cargo was out of her, the refit began. This was carried on night and day, the night gangs being lighted by the primitive means of torches and blazing tar barrels.

Plank tramways were laid from the railway depot to the ship's side, by which the huge Oregon masts and spars were brought to the water's edge. Using a big hermaphrodite derrick instead of mast sheers, Nicholas soon had the lower masts on end, after which re-rigging was carried on with a rush.

New sails, standing and running rigging were made by the original contractors in record time, so there was no waiting; yet, though all hands worked with a will, an unparalleled series of obstacles fought to delay the refit.

Snow blizzards swept the land, followed by intense frost—such weather as had not been experienced in Cornwall for 50 years. Then the imported riggers caused trouble, through the jealousy of the local men, and this resulted in riots and bloodshed. Yet, in spite of a foot of snow on the level and constant rows between the warring factions, the work went on like magic.

The new rigging was dropped over the mastheads, the yards crossed and sails bent, the hull itself repaired and remetalled by Falmouth shipwrights—who were more plentiful then than they are now—the boats repaired or new ones built, and lastly the cargo and stores safely stowed—all in six weeks. And on 31st January, *Sir Lancelot* once more set sail for China. The rest of the passage was uneventful, the new Oregon pine masts were a great success and Robinson swore by them. Though too late for the Foochow teas, the *Sir Lancelot* was able to sail from Shanghai on 15th June.

"Ariel's" Record Passage out to Hong Kong in 1866-7.

Whilst *Titania* was refitting in Rio and *Sir Lancelot* in Falmouth, *Ariel*, the first of the celebrated Steele trio, was covering herself with more glory.

Leaving Gravesend on 14th October, she arrived in Hong Kong harbour on 6th January, 1867, after a record passage of 79 days 21 hours, pilot to pilot, or 83 days from Gravesend to Hong Kong, anchorage to anchorage.

This wonderful passage made against the N.E. monsoon raised quite a sensation in Hong Kong, and when it was telegraphed home was hardly believed. It was an easy record for the run out to Hong Kong and has never been beaten since. There were many imaginative reports of better performances, there was even a rumour that an American clipper, the *Pride of the Ocean*, had run from the Lizard to Hong Kong in 69 days, but this, like all the others, was never substantiated. The nearest approaches to *Ariel's* passages that I can find are two runs of the *Northfleet*, both just under 90 days, and one of *Robin Hood's* of 90 days.

As Captain Keay wrote in his abstracts at the time:—"There were many reports of quicker passages than ours talked of by lovers of the marvellous, but on best authority in Hong Kong there was found to be no foundation for the mythical things said to have been done by sone gun-brig or by some clipper. Several naval officers visited us for a look at our chart and track out, also surveyors of long experience in China, and all agreed as to its being the fastest on record by some five or six days in any season, hence very difficult to beat in the N.E. season."

The Tea Race of 1867.

Owing partly to the dead heat finish in 1866, but perhaps more to the slump in tea, the 10s. per ton premium which amounted to about £500, was withdrawn in 1867; but for all that the racing continued as keen as ever. With the abolition of the premium it was arranged that the vessel making the best time was to be considered the winner and not the first in dock as heretofore.

The passages this year were as follows:—

Ship	Captain	From	Date Left	Anjer	Date of Arrival	Days Out
Taewan	—	Foochow	May 30	—	—	—
Maitland	Coulson	,,	June 1	—	Sept. 24	116
Serica	Innes	,,	,, 2	—	,, 30	120
Taeping	Dowdy	,,	,, 4	June 27	,, 14	102
Fiery Cross ..	Kirkup	,,	,, 5	July 1	,, 24	111
White Adder ..	—	,,	,, 6	,, 1	Oct. 7	123
Ziba	Jones	,,	,, 8	—	,, 7	121
Flying Spur ..	Ryrie	,,	,, 9	—	,, 2	115
Taitsing	Nutsford	,,	,, 9	July 14	,, 7	120
Black Prince ..	Inglis	,,	,, 10	,, 14	,, 7	119
Yangtze	Kemball	,,	,, 12	,, 14	,, 7	117
Ariel	Keay	,,	,, 13	,, 9	Sept. 23	102
Chinamen	—	,,	,, 14	,, 18	Oct. 7	115
Golden Spur ..	—	,,—	,, 18	—	,, 15	119
Deerfoot	—	Whampoa	,, 4	July 1	,, 7	125
Min..	Smith	,,	,, 7	,, 12	,, 8	123
Belted Will ..	Graham	,,	,, 7	,, 1	Sept. 24	109
John R. Worcester	—	Shanghai	,, 10	—	Oct. 12	124
Eliza Shaw ..	—	,,	,, 14	—	,, 11	119
Challenger	Brown	,,	,, 14	Aug. 12	,, 19	127
Sir Lancelot ..	Robinson	,,	,, 15	—	Sept. 22	99
Falcon	—	,,	July 8	—	Oct. 31	116
Titania	Deas	,,	Sept. 2	Oct. 9	Dec. 26	115

The first two starters from Foochow were new ships. *Taewan* had the misfortune to be wrecked on the day of sailing and became a total loss; but *Maitland* had a splendid start, as she found a fine N.E. breeze outside and was able to stretch away south with her moonsails set.

This vessel was expected to make a great reputation for herself. Her captain had boasted that she had run 17 knots an hour on the passage out and she had an unusually large sail spread, but she failed to come up to the expectations of her owners, who had asked Pile of Sunderland for a world beater. But speed is always elusive and though *Maitland* was fast enough off the wind, she would not go to windward like the Steele cracks.

Following the two new ships came the three old rivals, *Fiery Cross*, *Serica* and *Taeping*, the two latter with the same skippers as in 1866, but *Fiery Cross* was commanded by Captain Kirkup, Robinson having gone to the *Sir Lancelot.*

Ariel met all three ships on her way up to load. She passed *Serica* on the afternoon of the 3rd, the wind being very light from S.W. by S. *Taeping* towed past her in the outer channel on the afternoon of the 4th, and she had to wait at Sharp Peak till 7.30 p.m. on the 5th for the return of the tug *Woosung*, which had left Sharp Peak at 9 a.m. with *Fiery Cross* in tow.

The *Taitsing* and the veteran *Flying Spur* had a great race as to which should be loaded first; in the end *Flying Spur* just managed to finish an hour ahead, so had first call on the *Woosung*. No other tug being available, Captain Nutsford of the *Taitsing*, who considered that his vessel was much the faster of the two, tried hard to get Captain Ryrie to allow the Woosung to take him to sea first, but the latter naturally refused, not wishing to lose 24 hours. However, *Taitsing's* skipper was not to be beat and engaged fifty large sampans to tow his vessel down and by their means was enabled to get across the bar only an hour or so after *Flying Spur*. *Black Prince*, the whipper in, was only a tide behind them. Outside, the weather being hazy, the vessels were not long in sight of each other, but there was not much to choose between *Flying Spur*, *Taitsing* and *Black Prince* in light weather and they were never very far apart all the way down the China Seas.

On 13th June the *Flying Spur* was jogging along off the Cochin China Coast under short sail, having just experienced heavy weather when the *Taitsing* was sighted coming up astern under every rag she could set. Captain Ryrie at once made sail, and though the *Taitsing* occasionally crept up to *Flying Spur*, she never succeeded in passing her, and more than once the latter ran away from her and left her out of sight astern.

In the Api Passage *Flying Spur* and *Black Prince* found themselves in company. A heavy squall making up struck the two ships simultaneously. The careful Inglis at once clewed up his topgallant sails and light sails and kept away before it, but Ryrie luffed through it and when it cleared up the crew of *Black Prince* had the mortification of seeing the *Flying Spur* far away to windward sheeting home her royals. Of the three the *Flying Spur* was first through the Straits of Sunda.

The *Black Prince* nobly upheld her title of the "whipper in" by laying her main topsail to the mast and indulging in a few hours' fishing on the Agulhas Bank; nevertheless she managed to round the Cape in company with *Taitsing* on 13th August.

The *Flying Spur* passed the Cape two or three days earlier. Here she fell in with the *Sir Lancelot*, which had only sailed from Shanghai on 16th June, but under the energetic Robinson had made a marvellously quick run south.

It was a stormy day when the *Flying Spur* and *Sir Lancelot* met off the Cape Coast, and the wind was a "dead muzzler"; *Flying Spur* was carrying what was considered by her officers to be a heavy press of sail, viz., whole topsails and courses with outer jib, whilst other ships in company were close reefed. But *Sir Lancelot*, coming up on the opposite tack so as to cross the others' bow, was actually carrying three topgallant sails and flying jib.

Captain Ryrie and his officers looked at the approaching clipper with amazement, for the amount of canvas *Sir Lancelot* was staggering under was tremendous considering the wind. Indeed such cracking on would not have been possible but that the "Flying Horse" clipper had the run of the sea abaft the beam, whereas *Flying Spur* and the ships on the other tack had it before the beam.

As the two clippers converged on each other, they began signalling, and this nearly led to disaster on the *Sir Lancelot*, for just as she was athwart the hawse of the *Flying Spur*, her helmsman paying more attention to the latter's signal halliards than to his own steering, allowed his ship to come up in the wind and get aback. In a moment the *Sir Lancelot* had heeled right over and, getting sternway, was within an ace of being dismasted. Indeed, so far over did she go and so close were the two ships to each other, that the crew of the *Flying Spur* could see everything that took place on her decks. They saw Captain Robinson spring upon his careless helmsman, knock him down and jump on him, and they saw the watch below come flying on deck in their shirt tails. However, *Sir Lancelot's* crew were as smart as paint in whipping the sail off her, and the gallant clipper, as soon as she was relieved of some of the pressure aloft, brought her spars to windward and stood up, but it had been a close shave.

After this exciting episode the two ships were in company for ten days, a proof that *Flying Spur* when hard sailed could see the way in fair winds with any of her newer sisters.

Running down to St. Helena, the *Sir Lancelot* and *Flying Spur* both overhauled the moonsail clipper *Maitland*, and their treatment of her must have been more than trying to Captain Coulson after his bragging on the China Coast.

When the *Sir Lancelot* passed her, *Maitland* had every kite in her well-filled sail-locker hung out, from moonsails and skysail stunsails to watersails and save-alls. Captain Robinson had the usual flag talk as he was passing, and then as *Sir Lancelot* quickly drew ahead, he signalled sarcastically—"Goodbye, I shall be forced to leave you if you cannot make more sail."

Captain Ryrie of the *Flying Spur* was even more contemptuous. The *Flying Spur* sighted the wonder right ahead and coming up with her very fast went by her to windward, then, crossing over to the *Maitlands*' lee bow, Captain Ryrie backed his mainyards and let the *Maitland* pass him again, then refilling, he again weathered on her and again sailed past her to windward without any difficulty. This is the manoeuvre which is the origin of the term "sailing round a vessel," and it is undoubtedly the most humiliating dressing down that one vessel can give another. This experience, coming right on top of Robinson's sarcasm, must have taken most of the starch out of Captain Coulson's braggart spirit.

The winds this year were abnormally strong in the Atlantic, especially the S.E. trades, which were blowing an easterly gale off Ascension when *Black Prince* passed the island.

When in 12° N. and expecting the N.E. trades, *Black Prince*, after a squally, variable night, was caught aback at 4 a.m. by a sudden burst of wind from the norrard, which was strong enough to make her furl topgallant sails. At daylight the wind came away south again with a most unaccountably nasty sea and thick weather. All that morning she was passing through a fleet of outward bounders carrying very low sail, some of them with only a main topsail set.

At 10 a.m. a vessel was descried steering north under a couple of topsails and foresail; but on getting a better sight of her, those on the *Black Prince* saw that her jibboom was gone and that her topgallant masts with all their gear were lying across the backstays.

By 11 o'clock the Baring clipper was alongside the cripple, which proved to be her rival *Taitsing*.

It appeared that at 4 a.m. when *Black Prince* was caught aback in an ordinary squall, *Taitsing*, 20 miles to the north of her, was dismasted by a tornado of great violence, which accounted for the low sail carried by the outward bounders and the choppy sea. Captain Inglis offered assistance, but none being required the *Black Prince* went ahead and soon dropped *Taitsing* below the horizon.

Nutsford soon had his clipper re-rigged and a day or so later fell in with the *Flying Spur*, and both vessels were some days in company in doldrum weather on the edge of the N.E. trades, *Black Prince* being somewhere just behind the horizon.

The tornado had evidently upset the weather, but when the trades did come along, *Flying Spur* went ahead and eventually arrived in London five days ahead of her rivals.

We will now return to *Ariel*, which only arrived at the Pagoda Anchorage on 6th June, and was the last starter but two from Foochow.

After her record outward passage she had been sent up to Yokohama against the N.E. monsoon, and then on her return to Hong Kong found orders to proceed to Saigon; this made her late in arriving at Foochow.

She finished loading her tea at 9 a.m. on 12th June and proceeded down the river in tow of the screw steamer *Undine*. At 9.30 a.m. on 13th June the pilot was dropped, the steamer cast off and she headed away S.S.E. under all possible sail. At first she had light, baffling southerly winds, but on 18th June she ran into a heavy S.S.E. gale which lasted with very high sea until noon of the 19th, when light southerly airs and calms set in again.

On 27th June, *Ariel* caught up the first of the racers which had sailed before her, the *Black Prince*, which she signalled. This vessel she dropped astern and to leeward without any difficulty.

On the night of the 29th *Ariel* had a narrow escape of being run down by the unfortunate *Titania* as I have already related, and from this date she had the usual succession of faint airs and calms as far as Anjer, which she passed at 9.30 a.m. on 9th July, 26 days out. This was by no means a good start. *Taeping* had passed Anjer 13 days before on 26th June, only 22 days out, which was the fastest time made that season.

Ariel, though not making such good times as on the previous homeward passage, continued to catch up and pass the ships which had started ahead of her.

On 10th July, with a fine S.S.E. trade, she sighted a tea ship at daybreak on her starboard bow. This turned out to be *Serica*, which had sailed 11 days ahead. At 11 p.m. that night *Ariel* was up with her and so close that Captain Keay hailed but got no answer.

"Must be unwell or in the sulks at our beating him," was Keay's comment in his journal.

All the next day *Serica* continued to drop astern at about half a mile an hour, until at 5 p.m. her royals were only just visible above the horizon.

Three days later *Ariel* passed *White Adder*, sighting her at 7 a.m. and signalling her at noon; and at 7 a.m. on 15th July, just 24 hours after being sighted ahead, the Willis clipper disappeared below the horizon astern.

On 16th July *Ariel* made her best run—320 miles. She passed Mauritius on 23rd July and rounded the Cape on 8th August in an N.N.E. gale. She had taken 56 days from Foochow to the Cape, ten days longer than in 1866.

Owing to the strong trades, the ships had a better run up the Atlantic this year than in 1866, and if only *Ariel* had made as good a time to the Cape as she had done on her previous passage, she would have been home in 92 days. She passed west of St. Helena on 18th August, ten days from the Cape; and two days later overhauled *Belted Will*, which had been nine days ahead at Sunda Strait. At daylight on 20th August *Belted Will* was sighted ahead on *Ariel's* port bow, and at 10 a.m. on the 21st she went out of sight on the latter's starboard quarter.

On 25th August *Ariel* crossed the equator.

On Sunday, 8th September, at daylight *Ariel* sighted a ship right astern with a Jamie Green and other tea clipper kites set, which was thought to be *Fiery Cross.* However, this was a mistake as Captain Keay found out when he signalled *Fiery Cross* a week later. The wind was baffling and all round the compass, but in the afternoon the unknown racer astern dropped out of sight.

On 15th September *Ariel* came up with *Fiery Cross*, which was 3 miles ahead at daylight. The two ships exchanged signals and compared Greenwich time. The wind was fresh from E.S.E. with strong gusts, both vessels were braced sharp up on the starboard tack with royals and staysails fast, and it was soon evident that on this point of sailing there was little to choose between the two, *Ariel* only managing to weather and head reach on *Fiery Cross* a mile or so, though the former's run for the 24 hours was 260 miles. It took *Ariel* five days to shake off the wonderful veteran. On the 17th she had run *Fiery Cross's* hull down astern, but at 8 a.m. on the following day, the latter bobbed up again, passing 3 or 4 miles to leeward of *Ariel* on the opposite tack. After this Captain Keay saw her no more and eventually arrived 24 hours ahead.

On Friday, 20th September, at 6.30 p.m., *Ariel* made the Bishop Light, 43 days from the Cape and 99 from Foochow. The following is an abstract of her splendid run from the Cape to soundings:—

Passed St. Helena,	Aug.	18	..	10	days from	Cape Meridian
,, Ascension,	,,	21	..	3	,,	St. Helena
,, Equator,	,,	25	..	4	,,	Ascension
,, Cape Verd Isls.,	Sept.	1	..	7	,,	Equator
,, Flores, Azores,	,,	11	..	10	,,	Cape Verd Islands
,, Scillies,	,,	20	..	9	,,	Flores, Azores

This was eight days better than her 1866 time for the same run, but her passage had been spoilt by poor winds in the Indian Ocean.

Whilst *Ariel* and *Fiery Cross* were beating up for the Scillies against south-easterly winds, Captain Robinson was sending *Sir Lancelot* along a clean full and by. He was one of those skippers who did not mind additional mileage if he could only keep his ship moving, and he had a great objection to jamming the yards hard on the backstays if it could possibly be avoided. Thus, under him, *Sir Lancelot* was never pinched or jammed in the wind's eye. On 19th September he made the Mizen Head, 96 days out from Shanghai. Then, beating across from Ireland, passed to the norrard of the Scillies whilst *Ariel* was passing to the southard.

Captain Robinson's navigation on the night of the 20th was so daring as to scare his crew. According to the account given by *Sir Lancelot's* carpenter, there was great excitement amongst the men, and more than one grizzled old deep-water shell-back shook his head and asked, "Where the hell's the old man taking us?"

Passing between the Scillies and Land's End, Robinson beat to windward of the Seven Stones Lightship and the Wolf Rock. Luckily, it was a clear moonlight night and a smooth sea, the wind being light though ahead. And so successful was this piece of navigation that *Sir Lancelot*, cutting out *Ariel* by some hours, passed the Lizard before daylight on the 21st and went romping up Channel with a fine off-shore wind.

Ariel passed within 1½ miles of the Lizard Point at daylight, but was not so well favoured as to wind, which by that time had dropped away to a faint air.

All that day the two ships crowded sail up Channel, the wind being westerly and very light until noon, after which it freshened to a moderate breeze.

Sir Lancelot made Deal early on the 22nd, 99 days out from Shanghai; *Ariel*, not quite so well served by the wind, got her pilot just east of Dungeness at 2 p.m. on the same day, and towing all night with two tugs ahead, hauled into the East India Dock at 7 p.m. on the 23rd, exactly 102 days from Foochow. Though she had made the best passage from Foochow and beaten the *Taeping* by five hours on time, her success was overshadowed by the splendid performance of her sister ship, *Sir Lancelot.*

At last the enterprise of *Sir Lancelot's* owners was rewarded, and their vessel in one bound at the head of the racing fleet, her reputation made.

That she was really faster than either *Ariel* or *Titania* I do not believe, but one thing is certain, Captain Robinson was the sail carrier amongst a fleet of sail carriers and got the last ounce of speed out of his vessel.

Taeping, though she did not make the best time, had the satisfaction of getting her teas on the market a week before the next arrival. Captain M'Kinnon, the hero of 1866, had died on the voyage out, but his place was ably filled by Captain Dowdy.

It will be noticed that no less than seven of the tea ships docked in London on the same day, one of which was the *Yangtze*. This ship made the best time of the seven, though she was admitted to be the slowest of them all. And it was this performance which made Captain Kemball's reputation and brought him to the notice of Messrs. Thompson of the Aberdeen Clipper Line, and thus gained him the command of the celebrated *Thermopylae* in 1868. He was one of those skippers who was not afraid of a narrow rock-studded channel, and on one occasion took the *Yangtze* through Atlas Strait by moonlight when the wind was unfavourable for the Ombay Passage.

The Tea Clippers built in 1867—"Spindrift," "Lahloo," "Leander", and "Undine."

The success of the sister ships *Ariel* and *Sir Lancelot* was not long to remain undisputed, especially with such keen racing owners as Rodger and Findlay in the field.

During the year 1867 four new cracks were being built—*Spindrift* for Findlay, *Lahloo* for Rodger, *Leander* for Joseph Somes, and *Undine* for J. R. Kelso. *Spindrift* came from Connell's yard. In design she was somewhat of a new departure, especially in the way of more length, being 22 feet longer than *Sir Lancelot*, though their tonnage difference was only 13 tons. Indeed, she had more beams to length than any vessel launched for the tea trade since the famous *Lord of the Isles*, the proportions being:—

Lord of the Isles	6·89
Spindrift	6·17

In fact, *Spindrift* had more beams to length than any other clipper built of wood.

She also had one of the largest sail plans in the tea trade, and was nicknamed the "Giblet Pie," being "all legs and wings."

These two factors gave her great speed in reaching winds, and she may be reckoned as one of the fastest of all the tea clippers. Findlay, whose ambitions had not been realised in *Taitsing*, was delighted with his new ships and her premature end broke his heart.

Lahloo was a very different and less extreme vessel. Rodger, in spite of the success of *Ariel* and *Sir Lancelot*, still clung to the old single topsails, and *Lahloo* was an enlarged and improved *Taeping*.

Like all Steele's creations she was a very beautiful and taking vessel to look at, crossing the dainty skysail-yard at the main, a feature which, to my mind, always gave a full-rig ship a thorough-bred look.

She was launched on 23rd July, 1867, her builder's measure ments being:—

Length of keel and fore rake	190 feet
Breadth of beam	33 ,,
Depth of hold	20 ,,
Tonnage	$985\frac{83}{94}$,,

She was commanded by John Smith, one of those daring skippers who carried sail and was not afraid of a reef-studded passage.

Leander was designed by Bernard Waymouth the well-known naval architect and secretary of Lloyd's Register, and was built by Lawrie of Glasgow for the famous firm of Joseph Somes.

In design she was right up to date, and should have made a great reputation as she was undoubtedly an exceedingly fast ship. But she was unfortunate in having one of those captains who was too fond of his grog.

Her first passage out to China was an exceedingly fast one, 96 days to Shanghai; this seems to have turned her commander's head, and from the day of his arrival to the day he sailed he never missed an opportunity of celebrating it in the fashion of the coast.

One anecdote will suffice to show his way of life and the reason why *Leander* did not do as well as she should have done.

During the intermediate passages, the *Lord Macaulay* arrived at Foochow to load poles for Shanghai, and found *Leander* lying there already loaded, with the Blue Peter flying, whilst a big champagne luncheon was going on aboard of her.

The festive skipper of Somes' ship at once sent across to ask Captain Care to join his party, but the latter excused himself on the plea of having to go ashore and report his ship. But it was not easy to make the *Leander's* "old man" take no for an answer and Care only escaped a carouse by rowing ashore.

The days passed and *Leander* still lay all ready for sea, whilst daily beanfeasts took place aboard. At last the *Lord Macaulay* loaded and sailed, and still *Leander* lay to her anchors. Then when her skipper did at last think of moving, he was in such a reckless mood that he attempted to take his clipper through a channel in the river which was so narrow and rock strewn as to be only possible to shallow-draft junks. However it was a short cut and that was good enough for the dare-devil captain of the *Leander*.

As was only to be expected, he stuck on a ledge of rocks, and scraped a good deal of false keel and copper away in kedging his ship off again. Indeed it was only by sheer good fortune that *Leander* did not leave her bones on the reef for good and all. There was, of course, an enquiry, but the culprit contended that he had struck some sunken wreckage, which story the Court was good enough to swallow and therefore refrained from dealing with his ticket.

Leander was patched up in time to load at Shanghai, though not in time to race from Foochow with the other cracks.

Undine, another effort from the famous yard of Pile of Sunderland, was a fine fast little ship, and had the merit of being able to stand up without ballast and with a clean swept hold.

The Tea Race of 1868.

With four new ships and all the old cracks tuned up to the limit of efficiency, the tea race of 1868 was one of the most keenly contested of the whole series.

In the intermediate passages there were many opportunities of testing the ships against one another; and in these *Ariel*, especially, maintained her reputation. Leaving London on the 22nd October, 1867, she went out to Shanghai in 106 days, beating her old rival *Taeping* by five days. Again, on the coast in a passage from

The Tea Race of 1868.

Ship	Captain	Cargo	From	Date Sailed	Passed Anjer	Arrived	Days Out
Belted Will	—	934 496 lbs.	Whampoa	May 27	—	Sept. 7	103
Ariel	Keay	1,221,508 ,,	Foochow	,, 28	11 a.m. June 22	1 p.m. ,, 2	97
Taeping ..	McKinnon	1,165,508 ,,	,,	,, 28	,, 22	2 a.m. ,, 7	102
Sir Lancelot ..	Robinson	1,250,057 ,,	,,	,, 28	,, 22	Midnight ,, 2	98
Spindrift ..	Innes	1,306,836 ,,	,,	,, 29	6 a.m. ,, 23	Midnight ,, 2	97
Lahloo	Smith	1,231,397 ,,	,,	,, 30	10 a.m. ,, 23	7·30 p.m. ,, 7	101
Undine	—	1,088,398 ,,	Whampoa	,, 30	—	,, 11	104
Black Prince ..	Inglis	1,051,300 ,,	Foochow	,, 31	,, 28	,, 30	122
Serica	Innes	967,500 ,,	,,	June 1	,, 28	,, 21	113
Fiery Cross ..	Beckett	867,600 ,,	,,	,, 2	—	,, 30	120
Ziba	—	690,300 ,,	,,	,, 3	—	Oct. 8	127
Chinaman ..	—	884,800 ,,	,,	,, 4	—	Sept. 30	118
Yangtze ..	—	—	,,	,, 7	—	Oct. 7	122
Forward Ho ..	—	—	Shanghai	,, 11	—	,, 17	128
Titania ..	Deas	—	,,	,, 13	—	,, 17	126
Leander ..	Petherick	—	,,	,, 13	—	Sept. 30	109
Taitsing ..	Nutsford	—	,,	,, 15	—	Oct. 19	126
Min	—	—	Macao	,, 13	—	,, 23	132

Saigon to Hong Kong the two vessels had a very even match, their times, pilot to moorings being—

Ariel	4 days 20 hours
Taeping	4 days 23 hours

From Hong Kong they went on to Foochow to load tea, and once more *Ariel* beat *Taeping* by a few hours, and anchored off Pagoda Island on 15th May, 78 hours from Hong Kong harbour.

Spindrift and *Lahloo* both showed that they were worth reckoning with by their speed on rice passages up the coast. *Spindrift*, though, nearly finished her career by getting aground off Cape St. James, but was luckily refloated without damage.

The competition to be first ship away from Foochow was consequently very keen again this year, and the start was a good one, the first three ships, *Ariel*, *Taeping* and *Sir Lancelot*, crossing the bar together, with *Spindrift* one day, and *Lahloo* two days, behind.

At 2.25 p.m. on 28th May, the three leaders dropped their pilots outside the bar and made all plain sail on a wind, *Sir Lancelot* having a slight lead of *Ariel* and *Taeping* bringing up the rear. At 2.45 p.m. *Ariel* passed *Sir Lancelot*, after which the two ships were separated by Turnabout Island, *Ariel* taking the north side and *Sir Lancelot* the south. On 29th May the wind had dropped away to nothing, and the ships were scarcely steering. *Sir Lancelot* was ahead at midnight, but towards morning on the 30th *Ariel* passed her again and, with a very light N.E. wind, gradually left her astern. By 5 p.m. on 31st May *Ariel* had run *Sir Lancelot* out of sight, and the two ships saw no more of each other during the race.

Meanwhile *Taeping*, *Spindrift* and *Lahloo*, served with better winds, had been closing up on the leaders and on the 9th of June, when in 12° 36, N., 110° E., with a light S.S.W. breeze, *Ariel* discovered *Taeping* and *Spindrift* to windward of her. At the same time *Undine* showed up 10 miles to windward of the other three.

The *Spindrift* had come along with good winds whilst *Ariel* had had squalls and faint baffling airs and north-easterly sea on the 1st, 2nd and 3rd of June.

On Wednesday, the 10th of June, the four ships were still close together, and I find the following entry in Captain Keay's abstract:—

"1 p.m., in stays, was taken aback with a severe squall and rain in torrents—blew away fore and main topgallant sails and royals, flying jib and mizen staysail. *Spindrift* came up and passed us, but we seemed to gain on her directly our fore and main topgallant sails were set again. They then tacked and passed some 1½ miles more to leeward. Can beat her on a wind certain."

On the 11th and 12th of June very squally weather was experienced, and split sails were the order of the day. At 10 a.m. on the 11th *Ariel* weathered *Undine* by about 2 miles.

On the 13th of June another of the racers, *Lahloo*, appeared in sight to windward of the leader, *Ariel*, and in the afternoon Captain Keay remarks:—

"*Lahloo* same bearing, not so far to windward. Took in our main topgallant, royal and sky staysails and we seemed to gain on her."

On the 14th the Borneo Coast in sight, *Ariel* passed within hail of *Lahloo* and then steadily dropped her astern. However, *Lahloo* hung on to her great rival and could still be seen astern on the evening of the 15th.

The 18th of June found *Ariel* and *Spindrift* beating through the Api Passage together. At 3 p.m., with a light S.W. breeze, they weathered Api Point, the *Ariel* slowly gaining.

Both ships headed inshore towards sunset in order to catch the land breeze, but Captain Keay was the most daring, for whilst *Spindrift* went about when fairly close in for fear of getting ashore, *Ariel* kept on with the lead going on until midnight when, with only 9 fathoms under her, she caught the first breath of land wind and, going about, stole a march on *Spindrift*, whose port light could be seen as she lay becalmed in the offing. Captain Keay had run it close enough, for the next three casts of the lead gave 5, 4½ and 4 fathoms, but his daring paid, as the advantage gained over *Spindrift* gave him a lead of 19 hours at Anjer.

On 19th June, when on the line, *Ariel* sighted a clipper, supposed to be the *Belted Will*, 6 miles to windward, and another further off to the S.S.E., and at 7 p.m. on the 20th she hailed a ship bound the same way. but could not catch her reply. At 3 a.m. on 21st June, *Ariel* sighted Rotterdam Island on the port bow, hauled up south to within a mile, then steered along the east side of the Stolzes Channel within a mile of the shore under all sail and fetched right through. That night with a fresh S.W. breeze she cut right through the westernmost of the Thousand Islands Group, a most daring piece of navigation, which was rewarded by her being the first ship to pass Anjer, 25 days out, though *Sir Lancelot* was only a few hours behind.

Meanwhile *Black Prince* and *Serica* had been running as close a race as the leaders. *Serica* picked up the *Black Prince* off the Natunas, and showed the crew of the Baring clipper a superb sea picture as she crossed the latter's bows in a fresh breeze with every sail set to perfection, from Jimmy Green and jib-o'-jib to the ring-tail.

However the two ships proved to be very evenly matched and were constantly in company right down to Anjer. In the straits of Sunda *Serica* took the lead and went right away, a good example being given of how necessary it was to take advantage of every little flaw of wind.

The trade wind may usually be expected to come along off Krakatoa. *Serica* was hull down ahead of *Black Prince* at this point. A squall came off the island, *Serica* heeled to it and headed gaily away for Mauritius. It was the first breath of the trade, and she carried it right away and made a very quick run to the Cape.

Meanwhile the timid *Black Prince* put her helm up and ran away before it until the strength of the squall had eased, then when she did finally haul to the wind, she was too far to leeward, the breeze fell away and she was soon lying becalmed, whilst her crew had the chagrin of watching *Serica* sink sail after sail over the horizon. Then for a couple of days she lay in sight of Java Head in a clock calm, a very unusual experience at that season of the year. But the consequences of this waste of opportunity and excess of caution went still further than this, for *Serica* rounded the Cape before a series of westerly gales set in, which kept *Black Prince* dodging under reefed topsails for a week.

A curious accident happened to *Black Prince* on the run from Anjer to the Cape. She was struck by a swordfish. Attention was first drawn to the fact by the appearance of what seemed to be a bolt projecting from the side of the ship just abreast of the foremast and 8 inches below the water line. This proved to be the bony sword, which had pierced through the copper sheathing and teak planking and only been stopped by the iron frame. The fish had actually left 8 inches of its sword sticking in the side of the *Black Prince.*

We will now leave the *Black Prince* and return to the leaders, whom we had followed as far as Anjer.

In the run from Anjer to the Cape *Ariel* still maintained a slight lead of the others; crossing the trades she did the following fine performance:—

June 28, ...	330 miles
June 29, ...	315 miles
June 30, ...	314 miles

Her week's run was just under 2000 miles.

She passed Mauritius on the 4th of July, 12 days from Anjer, and on the 19th of July rounded the Cape in terrible weather.

On the 17th of July the wind rose to a gale, and on the afternoon of the 18th blew a hurricane from S.W. with a terrific cross sea, which was made all the worse by the Agulhas current. Though buried and swept fore and aft by every sea, the gallant little ship behaved nobly, shaking herself clear of the cataracts of water like a duck.

At 8 p.m. on 18th July the gale lulled right away to a calm, only to break out again an hour later more furious than ever. It caught *Ariel* under fore and mizen staysails, fore and main upper topsails and reefed foresail.

Notwithstanding the efforts of all hands, the foresail could not be made fast, and the main upper topsail, also, could only be hauled up as close as possible with the spilling lines and reef-tackles. whilst the stout storm staysail on the mizen stay split from head to foot and had to be secured. Indeed it was only after a battle of hours that the gaskets were at last put upon the fore upper topsail.

All this time the ship was running to the N.W. with the wind a point or two abaft the beam. The *Ariel* steered beautifully, and only one man was needed at the helm, though a lee wheel was kept handy in case the helmsman should be overpowered or swept from his post by the cascading seas. Some idea of the amount of water on deck may be obtained by noticing the damage done. The binnacle light was washed out; the long boat stove; side of midship house burst in; spare spars set adrift; a lanyard in the fore-rigging cut through by the fore-tack; hencoops and fowls washed overboard together with gratings and all kinds of gear; break of monkey poop and bulwarks stove and one of the ports gone, not to speak of minor casualties too numerous to mention.

On the 19th the strength of the wind began slowly to abate, but the sea grew still more mountainous and confused. However, by dusk the gale had taken off sufficiently to allow Captain Keay to make sail.

The Cape was passed as follows:—

1.	*Ariel*	July 19
2.	*Lahloo*	„ 19
3.	*Taeping*	„ 20
4.	*Sir Lancelot*	„ 20
5.	*Spindrift*	„ 20

Spindrift, making a more westerly course from the Cape, crossed the equator on 5th August, one day ahead of *Ariel* and *Lahloo*, and two days ahead of *Taeping*. On 5th and 6th August, *Ariel* had an interesting encounter with the Dundee clipper *Corona*, This was a notably fast ship of 1210 tons, very squarely rigged with huge lower stunsails set from the lower topsail yards.

Signals between the two ships were exchanged at 6 a.m. on 5th August, *Corona* being 63 days out from Bombay, homeward bound with troops and in good trim for showing her best paces. The S.E. trade was slowly dropping away, and in the light following wind the *Corona* held on well to the famous tea clipper, being still in sight on 6th August, 8 miles astern.

On the 12th of August *Spindrift*, *Ariel*, *Taeping*, and *Lahloo* were close together in doldrum weather.

Taeping sighted *Sprindrift* to the E.S.E. early in the morning, and later in the day fell in with *Lahloo* and *Ariel*, *Lahloo* being the windward and *Ariel* the leeward ship. On the following day at sunrise *Taeping* found herself 4 miles dead to windward of *Ariel*, and *Lahloo* was hull down on *Ariel's* weather beam, a light N.E. trade blowing. As the trade increased, *Ariel* gradually went ahead of *Taeping*, but the new flyer *Lahloo* reached and weathered on both of them. On 15th August *Ariel* had run *Taeping's* lower yards down on her lee quarter but *Lahloo* was out of sight to windward. At 7 a.m. *Taeping* bore S.W. ½ S. from *Ariel*, her main topsail yard dipping, but 4 hours later the wind *Ariel* was holding suddenly broke off to west, and *Taeping*, carrying the true wind, neared her rival by 2 or 3 miles. On the 17th at 6 a.m. *Taeping* was still to be seen, clinging to *Ariel's* skirts, hull down from the mizen crosstrees, but by 11, under the influence of a fine E.N.E. breeze, *Ariel* at last managed to sink her old antagonist below the horizon.

Ariel, the leader, made a fine run of 7 days from Cape Verd to the Western Isles, which she passed on 21st August, one day ahead of *Spindrift*. On August 28th *Taeping*, *Sir Lancelot*, *Spindrift*, and *Lahloo* were in company in 46° 10, N., 16° 48, W., *Ariel* being a day ahead in 47° 48, N., 13° 28, W. At 4 a.m. on 31st August, the Lizard lights were sighted from *Ariel's* topsail yard. She had a very light westerly wind for the run up Channel, but managed to get her pilot at Dungeness at 11.30 p.m. on 1st September, 96 days from pilot to pilot, and docked next day at 1 p.m. 97 days out. For the second time Captain Keay had brought his ship in first with the new season's teas.

But there was no close racing up Channel this year, the finish was as close as ever, for both *Spindrift* and *Sir Lancelot* docked the tide after *Ariel*, getting through the dock gates just before midnight on 2nd September. As *Spindrift* had sailed 23 hours after *Ariel*, she was considered the winner of the race. *Taeping* and *Lahloo* were very unlucky. *Lahloo* was becalmed 30 hours off the Eddystone, and *Taeping* flogged her sails against the masts off the Scillies during the whole of August 31st and September 1st.

Meanwhile far away behind the leaders, the old favourites *Serica* and *Fiery Cross* were not making such good passages. *Fiery Cross* especially had been caught like *Black Prince* by the unusual calm spell in Sunda Straits and the bad weather off the Cape. The two clippers met in 12° N., and Captain Inglis actually visited Captain Beckett of the *Fiery Cross*, nothing dismayed by a 4-mile pull in a blazing sun. After this the ships were together in doldrum weather for over a week.

At last after a heavy westerly gale the *Black Prince* found herself on soundings in the chops of the Channel. The wind was dead aft, but though it fell away in the night, no sail was made, and it was not until daylight that the cautious Inglis started to shake the reef out of the main topsail. The crew were aloft doing this when an outward bound ship came across her stern. This proved to be the *Ariel* plunging to the southward close on a wind, with only a single reef in her topsails. She actually had more canvas set on a wind than the *Black Prince* had running, and this incident was a standing joke against the leisurely methods of the "whipper in" for ever after.

The *Black Prince* arrived off the Ness on 29th September, and hove-to all night for a pilot. It was blowing hard and on the following day she again brought up for 4 or 5 hours in Margate Roads, nearly losing an anchor in the operation. And whilst the *Black Prince* was getting underweigh again, the *Fiery Cross* towed through, having beaten her in the run from Foochow by two days.

Thermopylae.

We have now come to the great *Thermopylae*, the pride of the British Merchant Service and justly considered by most seamen to have been the fastest sailing ship ever launched.

She was a more powerful ship than the dainty Steele clippers and had a good deal the best of even *Ariel*, *Sir Lancelot* or *Titania* in running the easting down; at the same time she was very fast in light airs. Indeed she has been known to have gone along 7 knots an hour when a man could have walked round the decks with a lighted candle. In fact, under every condition of wind she was a wonder. In steady quartering breezes, when all sail was set, she would go 12½ to 13 knots comfortably, her helm amidships and a small boy steering and there was never any necessity to take in her royals or small staysails until she was running well over 13 knots.

She went to windward like a witch and was equally good off the wind. With the wind right aft under foresail and fore lower topsail, main topsails, topgallant sail and royal she easily logged 13 and possibly the only vessel that could beat her in strong favouring winds was her great rival, the *Cutty Sark*.

Thermopylae was a splendid sea boat and when hove-to in bad weather rode out the worst sea like a duck, whilst she made good weather of it running under three lower topsails, reefed main upper topsails and reefed foresail.

Like all fine-lined ships she was wet enough when heavily pressed through a head sea, but with more bearing and less counter she did not scoop up the seas over her stern like such yacht-like vessels as *Ariel* or *Titania*.

Her designs compared with the half model of *Titania* plainly show this difference. *Thermopylae* also had a rocker false keel, which was supposed to help her to windward.

With regard to her sail plan she marked an advance in the direction of width of canvas as opposed to height. She had nothing above her royal yards, but these were tremendous spars. Her main royal was 19 feet deep, and it required four men to put the gaskets on this sail. Her mainyard was 80 feet long, and her mainsail had a drop of 40 feet.

She loaded 1000 tons of tea on a draft of 21 feet 6 inches with over 250 tons of ballast.

Kemball, who had previously commanded *Fairlight* and *Yangtze* was placed in command, and in starting under him *Thermopylae* was fortunate, for Messrs. Thompson could not have picked a keener or more enterprising skipper.

Thermopylae always called up admiration from every sailor who saw her. A small instance of this will suffice. On one occasion she cleared Port Phillip Heads in company with H.M.S. *Charybdis*. Both vessels crowded sail on the same course, but as soon as *Thermopylae* had her canvas set she began to draw rapidly away from the warship, in spite of all the latter's efforts to stay with her. At last, when *Thermopylae* had conclusively proved her superiority, the captain of the *Charybdis* could not restrain his admiration, and hoisted the following signal in the Mercantile Code as he rapidly dropped astern:—"Good bye. You are too much for us. You are the finest model of a ship I ever saw. It does my heart good to look at you."

Thermopylae was launched on 19th August, 1868, and sailed for Melbourne from Gravesend on 7th November. Her first voyage, in which she broke the record on each passage, has often been recorded in print. I give her abstract log in the appendix. On her first passage it will be noted that she

Left Gravesend 5 a.m. 7th November	
Passed the Lizard 8th November	1 day out
Crossed the Line 28th November	21 days out
Crossed meridian of Greenwich 13th December ..	36 ,,
Sighted Cape Otway, N. ½ W. 12 miles, 7th January	61 ,,
Anchored in Port Phillip 9th January	63 ,,

From pilot to pilot her passage was only 60 days, during which she made no less than nine runs of over 300 miles, the greatest being 330 with a strong quartering breeze.

Regarding this passage, an old Blackwall midshipman wrote as follows:—

"I was in Melbourne when *Thermopylae* came in, and, of course, went on board to have a look at the new marvel. She had immensely square yards, and most beautiful lines both fore and aft. Her apprentices told me her skipper had driven her all the way, carrying on tremendously; but her spars and rigging were new and of the best material and stood the severe strain in splendid fashion."

From Melbourne *Thermopylae* went up to Newcastle, N.S.W., where she loaded for Shanghai, and then made the passage across the Pacific from pilot to pilot in 28 days, another record. From Shanghai she went down to Foochow to load tea, with the golden cock at her masthead, which raised so much indignation amongst the crews of those ships which had already won the blue ribbon of the sea, as the Foochow tea race might justly be called. And one may imagine the excitement amongst the shippers when it was known that *Thermopylae* was to load tea.

"Windhover" and "Kaisow."

Besides the wonderful green clipper, two other tea ships were built in 1868, the *Windhover* and *Kaisow*.

The *Windhover* was a good wholesome ship, very like the *Forward Ho.* She proved to be fastest off the wind, though very good all round, being one of those vessels which did exceedingly well in any weather without being anything specially remarkable.

Kaisow, the Steele clipper of the year, had all the ghosting qualities of her predecessors, and thus excelled in the light weather of the China Seas, but she was a fuller lined vessel than most of Steele's ships and so was not quite in the first flight when homeward bound with tea.

The Tea Race of 1869.

Ship	Captain	From	Date Left	Passed Anjer	Date of Arrival	Days Out
Ariel	Courtenay	Foochow	June 30	—	Oct. 12	104
Leander	Petherick	,,	July 1	July 27	,, 12	103
Lahloo	Smith	,,	,, 2	—	,, 12	102
Thermopylae ..	Kemball	,,	,, 3	July 27	,, 2	91
Spindrift ..	Innes	,,	,, 4	—	,, 18	106
Taeping ..	Dowdy	,,	,, 9	—	,, 25	108
Ziba	—	,,	,, 12	—	Nov. 8	119
Sir Lancelot ..	Robinson	,,	,, 17	Aug. 7	Oct. 14	89
Kaisow ..	Anderson	,,	,, 18	Aug. 23	Nov. 8	113
Black Prince ..	Inglis	,,	,, 20	*via* Ombay	,, 16	119
Windhover ..	Nutsford	,,	,, 22	—	,, 8	109
Serica	Watts	,,	,, 25	—	,, 15	113
Falcon	Dunn	,,	,, 27	—	,, 15	111
Min	—	,,	Aug. 7	—	Dec. 16	131
Flying Spur ..	Beckett	,,	,, 25	—	,, 24	121
Undine ..	Scott	Shanghai	April 2	—	Aug. 2	122
Forward Ho ..	Hossack	,,	June 10	—	Oct. 2	114
Titania ..	Burgoyne	,,	,, 16	July 15	Sept. 22	98
Taitsing ..	—	,,	,, 21	—	Oct. 14	115
White Adder ..	Moore	,,	July 16	—	Nov. 9	116
Maitland ..	—	Whampoa	,, 21	Aug. 23	,, 8	110
Silver Eagle ..	—	,,	Aug. 12	—	Dec. 21	131
Yangtze	—	,,	,, 19	—	,, 24	127

In spite of the opening of the Suez Canal freights were up to £5 again in 1869, and Foochow was still the favourite port, though already there were signs of it being supplanted by Shanghai and Hankow.

Ariel severely felt the loss of Keay's clever handling, but it was only natural that her new master should not have understood her every mood as Captain Keay had done. However, once again the shippers favoured her, and she was the first ship of the season to get away from Foochow.

This year's racing showed better results than any other, no less than three ships breaking the record for the homeward run during the S.W. monsoon, namely, *Sir Lancelot* and *Thermopylae* from Foochow and *Titania* from Shanghai. The racing also was wonderfully close between the other ships. The first three starters from Foochow, *Ariel*, *Leander*, and *Lahloo*, sailing within 72 hours of each other, all arrived home on the same day. *Leander*, however, was severely handicapped by having too much ballast. She was so sharp that she loaded right down to her marks before she was full of tea, a most unprecedented circumstance, and it was said that she had to have some of her ballast taken out.

Thermopylae caught and passed her crossing from Anjer, nevertheless *Leander* kept with Holt's steamer, the *Achilles*, from Anjer to Mauritius.

But the chief interest in the Foochow race lay in the wonderful performances of *Thermopylae* and *Sir Lancelot*. A comparison of their passages gives the following:—

Points Passed	*Thermopylae*	Days Out	*Sir Lancelot*	Days Out
Sailed from Pagoda Anchorage	4 a.m. July 3		7 a.m. July 17	—
White Dogs N.N.E. 15 miles ..			July 18	—
Passed Anjer	6 a.m. July 28	25	Aug. 7	21
,, Mauritius	Aug. 9	37	—	—
Off the Buffalo River	,,		,, 28	42
Off Cape Agulhas	,, 21	49	Sept. 1	46
Passed St. Helena	(On W.) ,, 29	57	(sig'll'd) ,, 11	56
,, Ascension	Sept. 1	60	—	—
,, Equator	,, 6	65	—	—
,, Cape Verd	,, 12	71	—	—
,, Azores	,, 23	82	—	—
,, Lizard	,, 30	89	Oct. 10	85
,, Dungeness	5 p.m. Oct. 1	90	,, 12	87
	(fair wind)		foul baffling wind	
,, Gravesend	Oct. 2	91	2 p.m. Oct. 13	88
Docked	,, 2	91	,, 14	89

Captain Kemball, on getting outside the Min River, shaped a course to go "east about" as it was called—in other words through the Ombay Passage instead of by Sunda—but finding fresh winds in 126° E., he turned back into the China Sea and took the usual route *via* Anjer. When six days on the other side of Anjer, he overhauled and signalled *Leander*, which he spoke on the 3rd August and lost sight of astern on the 6th August.

The only other vessel which *Thermopylae* encountered on the passage was the auxiliary *Achilles*, which, sailing from Foochow on 18th July, passed the *Thermopylae* in the baffling weather of Sunda Strait, only to be caught and dropped astern on 17th August. *Thermopylae* was very lucky in only experiencing two days of wind without east in it all the way from Anjer to the Cape.

Like *Ariel* in 1867, *Sir Lancelot* made a late start owing to the length of her intermediate passages. She had left the East India Dock on 3rd October, 1868, and arrived at Hong Kong on 10th January, 1869, 99 days out.

The following is an epitome of her intermediate passages:—

Left Hong Kong, 27th January, 1869	
Arrived Bangkok, 5th February, 1869	9 days out
Left Bangkok, 3rd March, 1869	
Arrived Hong Kong 24th March, 1869	21 days out
Left Hong Kong, 10th April, 1869	
Arrived Saigon, 20th April, 1869	10 days out
Left Saigon, 5th May, 1869	
Arrived Yokohama, 26th May, 1869	21 days out
Left Yokohama, 14th June, 1869	
Arrived Foochow, 20th June, 1869	6 days out

Leaving Foochow on the 17th July, she was very lucky with her winds on the China Coast, being only 21 days to Anjer.

Crossing the trades the energetic Robinson took full advantage of his splendid start and averaged 300 miles a day for a whole week with the trade fresh on the beam, *Sir Lancelot's* best run during this portion of the passage being 336 miles, which gave her an average of 14 knots for the 24 hours.

Perhaps the chief incident of the passage was the deceiving of *Spindrift*. This happened off the Cape. *Spindrift* was sighted on *Sir Lancelot's* port beam, and Captain Robinson, surmising that Captain Innes would never believe that the vessel in sight was the *Sir Lancelot,* which had sailed 12 days later than his own flyer, determined to try and keep his identity secret and thus signalled the number of the *City of Dunedin* instead of his own. Innes was completely tricked, for on his arrival he reported speaking the *City of Dunedin* off the Cape.

Sir Lancelot's record would have been even better if she had not met with light baffling winds in the Channel, so that she took four days from the Lizard to dock, compared with *Thermopylae's* two days.

After meeting her unknown antagonist on 31st August, *Spindrift* did not allow her to get very far ahead and eventually arrived four days only behind her.

The other new cracks, *Windhover* and *Kaisow,* were not equal to rivalling *Thermopylae's* wonderful record, and their performances in the race home were distinctly disappointing.

The race from Shanghai gave *Titania* her first chance of showing what she could do. This voyage Captain Deas had retired and been succeeded by Captain Burgoyne, a far more enterprising and energetic skipper, who soon showed that hitherto *Titania* had not had justice done to her. Making allowance for the extra distance, her 98 days from Shanghai was very nearly as good as *Thermopylae's* 91 from Foochow. And her 26 days' run between Shanghai and Anjer was as good as *Sir Lancelot's* 21 from Foochow. I say 26 days, as she was held up near the Saddles until 20th June, and Anjer Lighthouse bore N.E. 6 miles on 15th July.

"Cutty Sark."

Whilst *Thermopylae, Sir Lancelot* and *Titania* were breaking records, a very notable vessel was being built at Dumbarton. This was the world-famous *Cutty Sark,* the only vessel which could seriously dispute *Thermopylae's* contention that she was the fastest ship in the world.

In her design the *Cutty Sark* is one of the most interesting ships now afloat, for in her model the past is linked up with the present, the days of the Napoleonic wars with the days of wireless and the flying machine.

Just as the Baltimore clippers owed their model to the clever draughtsmanship of some dead and gone French naval architect whose work, seen in the beautiful lines of some old Republican privateer, was thus perpetuated by the knowing Americans, so on the opposite side of the world in Bombay Harbour, the hulk of a French frigate, renowned in her time for speed, gave her form to one of the fastest clippers the world has ever seen, and added a further testimonial to the skill of the old French designers. Indeed, one may safely affirm without fear of contradiction that, in the gradual evolution of design and improvement in build, the sailing ship owes as much, if not more, to the French draughtsman's cleverness than to that of either the British or American.

The link in the chain between the old French frigate and the *Cutty Sark* was Captain John Willis's famous clipper ship the *Tweed*, whose history is worth recording.

In the year 1854 the Parsees built a paddle-wheel frigate in Bombay Dockyard for the East India Marine.

This was the *Punjaub*, one of the most costly vessels ever constructed of wood, as she was built entirely of the finest Malabar teak. Her moulds were made from the drawings of Oliver Lang, who, in turn, drew his inspirition from the aforesaid French frigate.

The *Punjaub*, after distinguishing herself in the Persian War and Indian Mutiny, was sent home in the spring of 1862 to be sold. Clever old John Willis bought her, converted her into a sailing ship, and, placing her under one of his most trusted captains, W. Stuart, late of the *Lammermuir*, sent her out to see what she could do as a peaceful trader after her stirring years of warfare.

The *Tweed* measured, when converted:—

1745 tons register	39 feet 6 inches beam
285 feet length over all	25 feet depth
250 feet length registered	

with a poop of 66 feet long, and a foc's'le of 57 feet.

On her first voyage as a sailing ship the *Tweed* took out and helped to lay the telegraph cable in the Persian Gulf, being the largest sailing ship ever handled in the Persian Gulf at that date, and one of the very few windjammers ever employed in cable laying. She did the passage out to Bombay with the cable on board in

77 days, then spent many roasting months in the Gulf laying it down, and when she finally reached Bombay, her arduous task accomplished, half her crew were laid up with fever.

In Bombay Dockyard she underwent a thorough overhaul, her cabins especially being enlarged and refurnished so as to provide accommodation for officers when she was taken up by the Government to carry troops, a service for which she proved to be very well fitted. From Bombay she went to Vingorla, took the Seaforth Highlanders on board, and made the run home in 78 days. Her reputation as an exceptionally speedy ship was now made, and for some years after the Suez Canal had been opened the Government employed her to carry invalided troops home round the Cape, in which service she made some wonderful runs.

In the great Indian famine of the seventies she was chartered to carry rice between Rangoon and Madras, and here again her "wonderful dashes across the Gulf of Bengal" worthily upheld her reputation.

Then when she came to the Colonial emigrant trade her records were equally good, the best being:—London to Port Chalmers 69 days, Sydney to London 69 days. Whilst on the China Coast, Mr. Joseph Conrad relates that "In the middle sixties she had beaten by a day and a half the steam mail boat from Hong Kong to Singapore," and that officers of men-of-war used to come on board to take the exact dimensions of her sail plan and note the placing of her masts.

It must be confessed, however, that she owed a great deal to Captain Stuart. This man was as remarkable as his ship, and if the old *Tweed* was a marvel in light winds, her captain was a marvel in the "roaring forties." How he drove her! But she bore driving like the thoroughbred that she was. Mr. Joseph Conrad, who sailed with Captain Stuart in the *Tweed's* successor, an iron Glasgow clipper gives a delightful sketch of the dare-devil Scotsman in his *Mirror of the Sea*.

"He seemed constitutionally incapable of ordering one of his officers to shorten sail," says Mr. Conrad. "If I had the watch from eight till midnight, he would leave the deck about nine with the words, 'Don't take any sail off her.' Then on the point of disappearing down the companionway he would add curtly, 'Don't carry anything away.' "

Mr. Conrad thus describes the *Tweed's* commander in a letter to me:—

"Captain Stuart was already very grey in my time, but there were no other signs of age about him. He resembled strikingly the portraits of the famous Hobart Pasha; the same line of the brow, the same nose broadening at the end, the same short horse-shoe beard. Captain Stuart was a native of Peterhead, with Viking blood in his veins. He spoke without any accent and prided himself on the purity of his English. He was not the conventional sea dog at all; but he was a perfect master of sea craft in all its branches, a first-rate seaman, a born commander, and a smart business man. In fact, an accomplished shipmaster of the time when shipmasters were not hung on the end of a telegraph cable and had the whole conduct of their ships in their own hands. Old Willis had an unbounded confidence in him. He used to take the *Tweed* out for three years, practically without instructions, and the ship earned a small fortune under his command."

I have seen some of Captain Stuart's beautifully written and clearly expressed business letters, which, with an abstract log of the passage, came home to Old John on the *Tweed's* arrival in port, and their contents bore out Mr. Conrad's testimony as to his business powers. Mr. Conrad also bears witness to "his extraordinary gift of incisive criticism," and his story of Stuart's dilemma in the *Loch Etive*, when his mate, a man too deaf to "tell how much wind there was," continually over carried sail, is a vivid little character study of the *Tweed's* commander.

In 1878 Captain Stuart gave the *Tweed* over to Captain Bice and took the command of the *Loch Etive*. And strange to say, the *Tweed* refused to sail for her new master, who had been mate in her for years. Though a fine seaman and smart officer, as any man trained under Stuart was bound to be, the *Tweed's* new captain utterly failed to make the old ship show her former phenomenal speed, and the worry of it broke his heart, and before the voyage was out he died at sea from sheer vexation of spirit.

After ten years in the Australian and Calcutta trade the *Tweed* ran into a gale off Algoa Bay when bound from Cochin to New York with general cargo, and on the 18th of July, 1888, she lost all her spars overboard except the fore and mizen lower masts. She was picked up by the s.s. *Venice* and towed into Algoa Bay, but proved to be not worth her salvage and was eventually broken up.

The *Tweed* was old Willis's favourite ship, and was certainly one of the most remarkable ships of her time. It is, therefore, not surprising that Willis hoped to lower *Thermopylae's* colours with a vessel designed on the same lines, and so commissioned Hercules Linton to take them off and use them on the design of his new tea clipper, *Cutty Sark*, whose name, I need hardly mention, was taken out of Burns's most famous poem. Tam o' Shanter's beautiful witch, Nannie, with her long hair and cutty sark flowing in the wind, formed the figurehead of the new clipper, and at the same time danced, as a dog-vane, at her main truck, and still dances there in spite of over 40 years' buffetting by wind, scorching by sun, and drenching by rain. As a work of art this original figurehead rivalled the splendid Leonidas of *Thermopylae* and the armed knight of *Sir Lancelot*.

Cutty Sark had rather an unlucky start to her long and victorious career. Messrs. Scott & Linton, who received the contract to build her, failed before she was completed, and she was finished off by Denny Bros. This, naturally, occasioned some delay, with the result that in the hurry to get her ready in time to reach China for the first season's teas some of the work was rather scamped, and a good deal of her iron work aloft, such as the trusses and slings of the yards, gave way on her first voyage and had to be renewed. And this hurried completion may have had something to do with the losing of her rudder during the great race with *Thermopylae* in 1872.

In appearance *Cutty Sark* was a thoroughbred from truck to keel, yet she cannot be said to have borne any particular resemblance to either the famous Steele clippers or *Thermopylae*. She was a more powerful vessel in every way than the dainty Steele creations and more powerful, in my opinion, than even *Thermopylae*, for no clipper of her size has ever rivalled her performances in easting weather.

She was specially built to beat *Thermopylae.* The following measurements, taken from Lloyd's Register, prove that *Cutty Sark,* though practically of the same size as *Thermopylae,* yet was a sharper ended ship.

	Cutty Sark	*Thermopylae*
Character	19 years A1	19 years A1
Length	212 feet 5 inches	212 feet
Beam	36 feet	36 feet
Depth	21 feet	21 feet
Depth moulded	22 feet 5 inches	23 feet 2 inches
Tonnage (under deck)	892 tons	927 tons
Tonnage (gross)	963 tons	991 tons
Tonnage (net)	921 tons	948 tons
Raised quarter-deck	46 feet long	61 feet long

Though 5 inches longer than *Thermopylae, Cutty Sark* yet measured 27 tons less than *Thermopylae.**

In her sail plan she resembled the Steele clippers and crossed the dainty skysail yard at the main. She was, of course, a double topsail yard ship and her sail area was tremendous.

Like *Thermopylae,* she continued to carry her full suit of racing sails and spars for years after the older favourites had been cut down and in many cases converted into barques. But in 1880 her spars were reduced, 9 feet 6 inches being cut off her long lower masts and 7 feet off her lower yards, the other spars being shortened in proportion. Yet even with this great loss of sail area, she continued to make record passages and runs such as 182 knots in 12 hours and 67 days from Sydney to the Lizards under Captain Woodget, who had her until she was sold to the Portuguese. In the placing of her masts, she was evidently a smaller edition of the *Tweed,* the foremast of both ships being stepped much further aft than was usual at that date, at the same time her sticks had a good rake to them like the *Tweed's.*

With regard to her other points, an examination of the photograph will tell a sailor quite as much as a written description, although her portrait was taken after she had been cut down. The

* Since the issue of the earlier editions the author has received the sail plan of *Cutty Sark* (which is given as Appendix I) from Mr. John Rennie, by whom it was designed. Mr. Rennie was chief draughtsman with Messrs. Scott & Linton during the whole period of the firm's existence.—B.L.

beautiful proportions of her bow and stern will at once be noticed. The short transom and deepish counter fit the perfect curve of the stem to a nicety, and her poise on the water gives a delightful hint of birdlike buoyancy and liveliness.

In the matter of speed *Cutty Sark's* records were as numerous as *Thermopylae's*, though she was not, on the whole, so fortunate in either her winds or commanders, for when she had a hard-sailing skipper as in her first three voyages she was most unlucky with her winds, and such entries as the following are constantly to be met with in his abstract logs.—

"In all my sailing about the China Sea, I have never experienced such weather. The principal complaint all the way down has been want of wind with three days' exception."

"Very unsettled with unsteady winds, the strangest trade winds I have seen for some time."

"Sails clashing against the masts. If they do not get worn one way, they will another. During this voyage, for one day that they have been full, they have been two clashing about."

"The old music again, sails clashing against the masts. Very pleasant for people who are not in a hurry."

"*Cutty Sark* music again, viz., sails clashing against the masts."

"Strong gale from N.N.W. When the wind comes it is a head gale."

"I cannot rightly describe the weather but it is the strangest I have ever seen in this part of the ocean. Clouds passing to and fro from one quarter then from another, sometimes a squall, sometimes rain only, and the swell that rose yesterday afternoon from W.N.W. has gone round to north and is now coming flying up from N.E., sometimes nearly breaking. Wind, faint variable airs from north, south, east and west but never enough to give the ship steerage way. The ship is sometimes rearing nearly over end with this N.W. swell."

All these entries are taken from *Cutty Sark's* first passage home, and her exasperated skipper finally writes sarcastically, when still nine days from port:—

"We will have to say 'Weel done, *Cutty Sark*!' for this is certainly splendid work, 100 days from Shanghai. I sincerely hope our neighbours will have the winds we have had."

On her third voyage *Cutty Sark* was better served with wind, but just when she seemed assured of a splendid passage and a victory over *Thermopylae* her chances were spoilt by the loss of her rudder.

Captain Moodie left her at the end of this voyage, and the captains who succeeded him were not racing captains in the strict sense of the word and did not carry sail hard. However, in the Australian trade she was lucky in being under one of the hard old breed of sea dogs in Captain Woodget, who never lost a mile for want of sail.

In actual speed through the water, Captain Moodie measured her both by the common and patent log and found her going as much as 17½ knots an hour.

He gives her best 24 hours' run under him as 363 knots, which is the biggest run ever made by a tea clipper.

A letter appearing in *Fairplay* some years back, which ran as follows:—

> Sir,—In your last issue are some remarks on the record breaking of Captain Willis's famous *Cutty Sark*. Permit me to give my quota of evidence as to this ship's extraordinary performances. At a earlier period of her career she made 362 and 363 knots in 24 hours on two consecutive days.—I am, sir, yours, etc.,
>
> "One Who Knows."

As late as 1892 she did a day's work of 353 knots when running the easting down and covered 2180 miles in the week. In 1890 she ran 3737 miles in 13 consecutive days between the Cape of Good Hope and the Leeuwin. In 1889 she went from Green Cape to Sydney, a distance of 220 miles, in 17 hours, and on another occasion sailed 7678 miles in 30 days.

All these runs were made after she was cut down, so that the 363 miles under her full sail plan seems in no way an impossible feat.

Crossing the S.E. trades from Anjer, when homeward bound with tea she generally managed to make some big runs. The following table showing three consecutive days' work crossing from Anjer from *Cutty Sark's* 1872 homeward run, *Thermopylae's* record run in 1869 and *Ariel's* best in 1868, is of interest to show *Cutty Sark's* superiority in hard whole sail breezes. We may safely assume that all three ships were driven to the utmost; indeed Captain Keay's private log is one string of broken stunsail booms.

In comparing the speeds of *Cutty Sark* and *Thermopylae*, after a close study of their abstract logs for a number of voyages, I find myself drawing these conclusions:—

(1) That *Thermopylae*, though an exceptional ship in any weather, was at her best in moderate and light winds.

(2) That *Cutty Sark's* best point was with strong beam or quartering breezes, on which occasions she could beat *Thermopylae* or any other ship ever built.

(3) That *Cutty Sark* was the best ship to windward in rough water, *Thermopylae* in light winds and smooth water.

(4) That China voyages suited *Thermopylae* best and Australian voyages suited *Cutty Sark* best.

Ship		Miles	Remarks
Ariel, 1868			
June	28	330	S.E. and S.E. by E., strong
,,	29	315	S.E. fresh and strong gusts
,,	30	314	S.E. more moderate
Thermopylae 1869			
July	31	267	S.E. fresh
August	1	290	S.S.E. and strong
,,	2	318	E.S.E. strong
Cutty Sark 1872			
July	27	340	E.S.E. strong and squally
,,	28	327	E.S.E. more moderate
,,	29	320	E.S.E. moderate

Before leaving this much-disputed question of speed, I will quote from an old sailor's reminiscences in order to show *Cutty Sark's* speed in comparison with that of a fast main skysail-yard wool clipper in the year 1879:—

"One day we sighted a vessel, a mere speck on the horizon, astern of us, and the way she came into view it was evident she was travelling much faster than ourselves. Bringing the wind up with her was remarked on board, and that seemed the only feasible conclusion to arrive at and account for the manner in which she overhauled us. In a few hours she was alongside us, and proved to be the famous British clipper *Cutty Sark*, one of the fastest ships afloat. She passed us going 2 feet to our 1, and in a short time was hull down ahead of us."

Cutty Sark was launched in November, 1869, and the following is an epitome of her maiden passage:—

February 15, 1870—Made sail from the Downs about midnight.

February 16, 1870—Took departure from the Start at 11 p.m.

Had strong southerly winds and gales to the N.E. trades. Had very light N.E. trades.

March, 13, 1870—Crossed the line; 25 days out. Had very light S.E. trades.

April 6, 1870—Crossed the meridian of Greenwich; 49 days out.

Her best week's work running easting down was between 13th and 19th April, and totalled 2061 miles.

April 13, 1870—44° 18′ S., 38° 48′ E. Distance 298. Wind N.N.E. and north, strong.

April 14, 1870—44° 18′ S., 47° 8′ E. Distance 360. Wind north, fresh. P.M., moderate.

April 15, 1870—44° 25′ S., 53° 22′ E. Distance 269. Wind north, strong W.N.W. (Distance by common and patent log and dead reckoning 343 miles. Adverse current).

April 16, 1870—44° 9′ S., 60° 22′ E. Distance 304. Wind W.N.W. and N.N.W., strong to a gale.

April 17, 1870—43° 23′ S., 66° 21′ E. Distance 266. Wind north: light wind and heavy sea. P.M., fresh gale, passed a ship steering east.

April 18, 1870—42° 17′ S., 73° 51′ E. Distance 336. Wind N.N.E., fresh gale, Noon, S.S.W. Midnight, S.E. strong.

April 19, 1870—40° 27′ S., 78° 14′ E. Distance 228. Wind S.E., strong gale.

May 3, 1870—Made Java Head 76 days out.

In the Straits of Sunda *Cutty Sark* overhauled the Aberdeen White Star clipper *Thyatira*, *Doune Castle* (one of Skinner's), *Wylo* (the new Steele clipper), and the veteran *Fiery Cross*. With light variable airs and calms, these ships hung on to her longer than they would have done otherwise. *Wylo* proved the most difficult to shake off, but after being eight days in company she also at last dropped out of sight astern. *Cutty Sark* had light airs and calms all the way up the coast with the exception of three days of the N.E. monsoon blowing a gale from right ahead.

At 5 a.m. on 31st May she picked up her Shanghai pilot off the lightship, having, in spite of an unusual amount of light winds and calms, made the passage pilot to pilot in 104 days.

The following list of outward passages, made about the same time, show that *Cutty Sark's* run was a very good one.

Outward Passages to China 1869-70.

Ship	From	Left	To	Arrived	Days Out
Windhover	London	Dec. 4	Shanghai	May 1	148
Black Prince	,,	,, 18	Hong Kong	April 9	112
White Adder	,,	,, 18	Shanghai	May 21	154
Serica	,,	,, 20	,,	,, 10	141
Flying Spur	,,	Jan. 21	,,	,, 21	120
Thyatira	,,	Feb. 1	,,	June 1	120
John R. Worcester ..	,,	,, 3	,,	May 29	115
Fiery Cross	,,	,, 5	Hong Kong	,, 21	105
Wylo	,,	,, 11	Foochow	,, 25	103
Cutty Sark	,,	,, 15	Shanghai	,, 31	104

"Norman Court."

Next to *Cutty Sark* the most important clipper launched in 1869 was the *Norman Court.* Designed by Rennie, she bore a strong family likeness to *Fiery Cross* and *Black Prince*, and was a very beautiful ship in every way. She should have been Rennie's masterpiece, but the builders made some slight deviation from his design in the moulding of the iron frames, which, though it did not interfere very much with her speed, made her more tender than she should otherwise have been. This little deviation was necessary in order to bring Rennie's measurements within the Lloyd's scantlings for a thousand-ton ship.

However, with the exception of this alteration which affected her stability, *Norman Court* had beautifully fair lines, and she was most perfectly built and finished. Unlike *Cutty Sark's*, her iron work was specially good. In fact, a London blacksmith, who was employed repairing one of her trusses some years later, was so lost in admiration of her iron work that he declared it must have been made by a watchmaker.

As to her deck fittings, her bulwarks were panelled in teak, with a solid brass rail on top all round. And even her foc's'le lockers were panelled better than those of many a ship's cabin.

Norman Court, indeed, rivalled the Steele clippers in looks and beauty, and was considered at one time to be the prettiest rigged vessel sailing out of London.

She was very heavily sparred and extremely lofty, so lofty, indeed, that one 4th of July, when she was lying in Shanghai with several other clippers, including *Thermopylae*, the American superintendent of the Hankow Wharf came off with a star-spangled banner and asked Captain Shewan to fly it at his main truck, remarking that it would be seen further from there than from any other point within leagues of Shanghai. Captain Shewan was also asked whether he gave an apprentice a biscuit before he sent him up to furl the skysail. Indeed, if the Baring clipper had been as square as *Thermopylae* with her own loftiness she would have been very much overhatted, but, luckily for her stability, she had a narrow sail plan.

Like most of the tea clippers, her masts were raked well aft, in fact, they had more rake than was usual, and this, Captain Shewan thought, rather spoilt her sailing in light winds. The chief reason for this rake was that it kept a wooden ship from diving too much into a head sea.

In her paces *Norman Court* was a *bona fide* tea clipper in every way—fast in light airs, at her best with fresh whole sail beam winds but not the equal of *Cutty Sark* when the royals were fast, and perhaps a good ½ knot slower than the Willis crack when off the wind, for *Norman Court's* best point was to windward—indeed, she was one of the most weatherly of all the tea clippers. Owing to the way in which her bilge was carried right away to her stem (though there was nothing above the water line to stop her) she went into a sea like a rubber ball, and very rarely buried herself like some of the Aberdeen ships. She required careful watching, however, and if caught by the wind freeing two or three points in a squall when going close-hauled under a press of sail she would go over till the lee bunks of the midshiphouse were underwater.

With regard to trim, she sailed best, especially running, when well down by the stern. On one occasion, when she left London for Sydney with a light load-line, Captain Shewan kept her on an even keel, but found that she did not do as well as usual running the easting down. On the other hand, in 1871, when she made the fast run of 67 days to the South Cape, Tasmania, she was very deep with Manchester bales and nearly a foot by the stern. This trim gave her some splendid runs in the "roaring forties," but she also took a tremendous lot of heavy water over aft in making them. Once she left Macao in heavy weather with no chance to get her proper trim. This passage she sailed first rate on a wind, though very wet forward, and on her arrival she was found to be 6 inches by the head.

Norman Court was called after the Hampshire seat of her owner, Thomas Baring, and her figurehead was a splendidly-carved likeness of one of the family beauties. Captain Andrew Shewan, senior, had her for her first two voyages, and then his son had her for the rest of her racing career.

She was launched from Inglis's yard in July, and sailed for Hong Kong in October, making the very good passage out of 98 days.

The "Caliph."

Norman Court left on her maiden voyage in company with another new tea clipper. This was the *Caliph*, Hall of Aberdeen's starter in the race to displace *Thermopylae's* cock of victory.

Caliph, however, disappeared in the China Seas on her passage home. She loaded at Foochow for New York, and was one of the last of the tea fleet to sail. As there was no typhoon to account for her disappearance, it was supposed at the time that she had been surprised and captured at a calm by pirates, and from that day to this no trace of her has ever been found. However, though her life was so short she is worth a description, as she was a very up-to-date clipper in every way.

In design she was a very extreme ship with far more dead rise than any of her contemporaries. And she had as lofty a sail plan as *Norman Court*, crossing all three skysails. She also had height without width, and had much smaller courses and topsails than *Ariel* and *Sir Lancelot*, though she was of greater tonnage.

But the chief innovation on the *Caliph* was her engine. This was of 8-horse power, situated at the after end of her midship-house, as became the custom later in iron clippers. But, besides being useful for handling cargo, lifting the anchor, pumping ship, and hoisting topsails, this engine was fitted by means of shafts for driving two small screws, which could be lowered overboard on each side of the vessel, and were expected to drive the ship along 2½ knots in a calm. *Caliph* also had a condenser capable of distilling 50 gallons of water in 12 hours.

She was commanded by the son of a ship chandler or rope maker, who had some shares in her, a man without much experience, and this fact may have something to do with her disappearance.

"Wylo," "Ambassador," "Eme," and "Osaka."

Of the other clippers launched in 1869, *Wylo*, from Steele's yard, was a sister ship of *Kaisow*. *Ambassador*, Lund's first venture in the tea trade, was a very cranky and overmasted, though a fast ship. *Eme*, a very pretty vessel, was a light-wind flyer. She was, however, very unfortunate in her first race home and took 135 days to London. She went by the long eastern route and got badly becalmed. However, though it was more the misfortune than the fault of her skipper, old Captain Wade, of Robertson & Co., her owners, met the ship on arrival and sacked the wretched man there and then, with such a flow of language as made the neighbouring bargees gape with admiration. To be unlucky in the tea trade was a very bad offence, and no amount of good seamanship could make up for it in the eyes of the owners. The *Osaka* was a small clipper barque, built by Pile of Sunderland, for Captain Killick of *Challenger* fame, the owner of *Wylo* and *Kaisow*.

"Oberon."

Before turning to the tea race of 1870, I must say a few words about *Oberon*, the ship which Captain Keay left *Ariel* to take command of.

Oberon was an experiment of Maxton's, one of those failures as auxiliary steamers which, when their screws were removed, proved very fine sailing vessels. It is curious that this was by no means uncommon with early full-rigged steamers. The mention of the following names, the *Tweed*, *Oberon*, *Darling Downs*, *Lady Jocelyn*, and *Lancing* will show how successful the steamship design has been under sail.

Oberon was heavily rigged with three skysail yards, and when under sail alone proved a very fast, handy vessel and a splendid sea boat. But her best speed under steam, at a coal consumption of 7 tons a day, was only 7 knots an hour, and she proved to be quite unable to stem a strong head wind and sea.

She cost £35,000 to build, and thus was a very costly experiment.

Her black squad consisted of two engineers and three firemen. These poor wretches had a very bad time in the tropics, as in those days the ventilation of engine room and stokehole was most primitive.

Captain Keay, after a very worrying voyage, managed to get her home from Hankow *via* the Cape in 115 days. He then left her for another troublesome steam kettle.

Oberon's second voyage, with a less experienced captain, was an even greater disappointment than her first. She started well by making Port Said under sail alone in 18 days from Plymouth. The sails were then put in the gaskets and steam raised for the passage of the Canal and Red Sea. All went well until she was nearing the southern end of the Red Sea, when a strong southerly wind absolutely stopped her headway. Hoping that this would soon take off, the captain anchored off Mocha. But eight days passed and still the southerly wind blew as strong as ever, so *Oberon* was at last compelled to beat through Laage Strait under both steam and canvas. This soon consumed her limited supply of coal, and she was obliged to put into Aden to refill her bunkers. Coaling was again a necessity at Labuan.

Her next trouble was going up the Yangtze to Hankow. With great difficulty she managed to stem the 7-knot current as far as the Orphan Rock, but here the current became so fierce that her headway was completely stopped. Thereupon her engineer did what many another engineer was compelled to do in those early days of the steam engine, he jammed down the safety valve and raised the working pressure from 30 to 45 lbs. This desperate expedient just got her past the Orphan Rock in safety.

On the homeward run she took 134 days from Hankow *via* the Cape. Two such voyages were enough. Her machinery was removed, and henceforth she depended on sail alone with infinitely better results.

An important absentee will be noticed in the records of the 1870 homeward passages. This is the unfortunate *Spindrift*, which was wrecked on Dungeness when outward bound in charge of a Channel pilot, and the loss of this splendid clipper was most unsatisfactory and unnecessary.

The Tea Race of 1870.

Ship	Captain	From	Left	Passed Anjer	Arrived London	Days Out
Ambassador	Duggan	Foochow	July 25	Pitt's Pas'ge Aug. 24	Nov. 17	115
Thermopylae	Kemball	,,	,, 29	,, 23	,, 12	106
Eme	—	,,	,, 31	—	Dec. 13	135
Sir Lancelot ..	Edmonds	,,	Aug. 2	Sept. 4	Nov. 14	104
Norman Court	Shewan	,,	,, 3	,, 4	,, 16	105
Wylo	Brown	,,	,, 18	,, 27	Dec. 12	112
Chinaman ..	—	,,	,, 27	,, 25	,, 13	108
Windhover ..	—	,,	,, 30	—	,, 8	100
Falcon ..	Dunn	,,	Sept. 2	—	,, 19	108
Maitland ..	—	,,	,, 10	Oct. 6	,, 30	111
Flying Spur ..	Beckett	,,	,, 22	,, 21	Jan. 20	120
Lahloe ..	Smith	,,	Oct. 12	—	,, 18	98
Kaisow ..	Anderson	,,	,, 26	—	Feb. 2	99
Taitsing ..	—	,,	Nov. 3	—	Mar. 4	121
Taeping ..	Dowdy	Macao	June 10	—	Sept, 29	111
Belted Will ..	Thompson	,,	July 13	Aug. 13	Oct. 25	101
Fiery Cross ..	Middleton	,,	Sept. 16	—	Jan. 10	116
Oberon	John Keay	Shanghai	June 18	—	Oct. 7	111
Titania ..	Burgoyne	,,	,, 18	—	,, 8	112
Cutty Sark ..	Moodie	,,	,, 25	Aug. 2	,, 13	110
Serica	Sproule	,,	,, 28	—	,, 24	118
Forward Ho ..	Hossack	,,	,, 28	—	,, 25	119
Ethopian ..	Faulkner	,,	July 1	—	Nov. 12	134
J. R. Worcester	Cawse	,,	,, 5	—	,, 8	126
Undine ..	Scott	,,	,, 30	—	,, 14	107
Thyatira ..	M'Kay	,,	Aug. 12	—	Dec. 8	118
Leander ..	Petherick	,,	Oct. 12	—	Jan. 18	98

Captain Nutsford, who had only been home two days, was transferred into her from the *Windhover*, and with such a quick turn round decided to take his wife with him down Channel. The pilot also had only just returned from taking a ship out, and had had no time for any rest.

With a fine northerly wind and a clear night *Spindrift* was making splendid time, and in order to save mileage was hugging the Kent shore.

Captain Nutsford was asleep in the cabin and the pilot was dozing in the charthouse, when the second mate, who was in charge of the deck, roused the pilot with the news that the course he had set was taking her too close inshore. Apparently the pilot called out sleepily, "Luff! Luff!" which was at once obeyed by the helmsman, with the result that *Spindrift* slid right up on the beach with the lighthouse on her lee bow. And she had so much way on that it was found impossible to refloat her, and thus the beautiful ship became a total loss.

In the enquiry which followed both the pilot and captain escaped censure, though it was proved that there was an untried leadsman in the chains and that the pilot had mistaken a star for the Dungeness Light.

The best race this year from Foochow was that between *Thermopylae*, *Sir Lancelot* and *Norman Court*. On 12th November *Sir Lancelot* passed through the Downs as *Thermopylae* was docking, and at the same time *Norman Court* was reported off the Start. With a little luck the latter would have done the best passage of the three, but no sooner had the other two docked than N.E. winds set in and delayed her in the Channel.

The first starters from Shanghai were *Oberon*, *Titania* (both of which had loaded up at Hankow) and *Cutty Sark*.

Taking into consideration the abnormally light winds both in the China Seas and in the Atlantic, the times of *Titania* and *Cutty Sark* were very good. *Oberon* was rather unlucky striking such weather and had to steam the whole way from the N.E. trades to soundings, and was obliged to coal both at Cape Verde and Falmouth.

Captain Moodie drove *Cutty Sark* unmercifully in his efforts to make a notable passage in spite of a constant succession of light head winds, but he was not a daring navigator of the Keay type and was specially cautious in the China Seas.

The following is an abstract from *Cutty Sark's* log—

June 25—5 p.m., passed the lightship. Midnight, anchored in 7 fathoms; strong tide and no wind.

June 26—4 a.m., got underweigh, light southerly breezes. Amherst Rocks bearing N. by E. ½° E. 3 miles.

July 1—Tacked 6 miles off Tamsua Harbour (Formosa).

July 2—At 10 a.m., Turnabout Island, N. by W. 8 miles. Wind S.S.W., strong and short head sea. (From this point *Cutty Sark* beat doggedly down the centre of the China Sea to the Natunas against light south-westerly airs and a strong current.

July 23—At 8 a.m., tacked off south-east end of Great Natuna. (Captain Moodie, though favourably placed for going through the Api Passage and working the land breezes off the Borneo Coast, preferred to go between the Natunas.)

July 24—Blew a hard gale from west for two or three hours and *Cutty Sark* had to be shortened down to lower topsails and foresail.

July 25—7 p.m., passed Direction Island. 9.30 p.m., overhauled and passed a steamer. (Captain Moodie sacrificed three days going round by the Karamata Passage instead of cutting through Gaspar Strait.)

August 1—North Watcher W. by S. 6 miles.

August 2—9 a.m., a boat from Anjer came aboard—sent letters to be posted. Noon, St. Nicholas Point bore S. by E. 8 miles. (During the passage down the China Sea *Cutty Sark* had had to contend against 728 miles of north-easterly current.)

August 7 to 10—Daily runs were 303, 311, 320 and 314 miles. *Cutty Sark* had carried away so many stunsails booms that from the 7th she had to go for some days without stunsails.

August 16—Passed Mauritius.

August 28—Rounded the Cape of Good Hope in a strong gale from N.N.W.

September 8—At 9.30 a.m., saw St. Helena bearing N.N.W.

September 8-16—Very light S.E. trades.

September 12—10 p.m., passed Ascension 4 miles off.

September 16—Crossed the equator.

September 16-24—Doldrum weather.

September 24—Took N.E. trades in 14° 50′ N., 26° 12′ W.

September 28—Lost N.E. trades.

September 29 to October 5—Calms and light airs to 35° N., 33° W.

October 9—45° 28′ N., 18° 35′ W. Winds west to N.W. "Mostly strong but very unsteady both in force and direction. Ship is sometimes going over 14 knots, at others barely 10 knots. All port studding-sails set and the broken masts and yards stand pretty stiff to it."

October 10—Strong westerly gale with ugly sea; ship taking water at both ends at once. Run 290 miles.

October 12—At 9.30 p.m., Beachy Head N.N.E. 10 miles. Blowing hard from W.

October 13—At 7.30 a.m., anchored in the Downs, as it was blowing too hard to be able to tow.

Times between different Points:—

Shanghai to Anjer	.. .. ..	38 days
Anjer to Mauritius	.. .. ..	14 „
Mauritius to the Cape	.. .. ..	12 „
The Cape to St. Helena	.. ..	11 „
St. Helena to Ascension	.. ..	4 „
Ascension to the Equator	.. ..	4 „
Equator to Cape Verd	.. .. ..	9 „
Cape Verd to Western Isles	.. ..	11 „
Western Isles to Gravesend	.. ..	7 „
		110 days from Shanghai, 72 from Anjer.

It will be noticed that the best passages of the year were made by the late starters, who had the benefit of the favourable monsoon in the China Seas.

The mere fact that such heelers as *Leander* and *Lahloo* did not leave China until October shows that already steamers which took advantage of the Suez Canal, such as the *West Indian*, *Achilles*, *Nestor*, *Diomed*, *Agamemnon* and *Erl King*, were casting a shadow over the trade and giving more than a hint that the day of the racing clipper was coming to an end.

However, though the clippers were no longer the first to arrive with the new teas, they made a hard fight of it and still raced home without sparing canvas. Indeed, the racing owners were far from giving up the contest because the Canal was opened, and during 1870 three more notables ships were built to carry on the fierce battle between sail and steam.

The Unlucky "Black Adder."

The first of these was John Willis's *Black Adder*. A ship differs from every other work of man in one great particular. She has a soul, a living personality, which personality seems to be just as much under the influence of the Fates as that of any human being.

Some ships are lucky ships and bear a charmed life in which everything always goes right. Others—and any old seafarer could give one a string of names—seemed to be placed under an evil spell from the very day of their launch as if they had been born under an unlucky star.

Of these, perhaps the worst are the "mankillers," ships which, without ever running into trouble themselves, never arrive at the end of a passage without having lost men in some way or other, sometimes by being washed overboard, at others by falling from aloft or even by the wholesale scourge of some strange disease unknown to the medical profession.

Other ships gain a bad name for experiencing everlasting head winds, gales or calms; for collisions, strandings, fires, or running foul of ice; for a crooked disposition of some sort or other, such as breaking their sheer and losing anchors, refusing to manoeuvre or steer on occasions as if they had sudden fits of the sulks, or carrying away spars and losing sails without an apparent reason. But occasionally one comes across a ship which is unlucky in all these various ways and then one may truly speak of her as—

Built i' th' eclipse and rigged with curses dark.

And perhaps no ship ever deserved this description more than the tea clipper *Black Adder*.

So pleased was Captain John Willis with his wonderful *Tweed* that he was not content to build the *Cutty Sark* on her lines, but also must needs order two iron sister clippers to be built with the same under-water body as his favourite.

These were the *Black Adder* and *Hallowe'en*, the latter of which, strangely enough, was as lucky as the former was unlucky, though the two ships were built side by side on the London River by Messrs. Maudsley, Sons & Field, from lines taken off the *Tweed* by Messrs. Ritherdon and Thompson, the surveyors to the East India Council.

In appearance above the waterline they bore very little resemblance to *Cutty Sark*, and had the usual iron ship's topgallant foc's'le and short turtle-backed poop.

The following comparison of their measuresments will therefore be of interest—

Ship	Construction	Net Tons	Gross Tons	Under Deck Tonnage	Length	Beam	Depth
Cutty Sark ..	Comp.	921	963	892	212·5	36	21
Black Adder	Iron	917	970	872	216·6	35·2	20·5
Hallowe'en ..	Iron	920	971	873	216·6	35·2	20·5

Unfortunately, Maudsley & Co., at that time had had more experience in marine engineering than in actual shipbuilding, and this fact had undoubtedly a good deal to do with *Black Adder's* early misfortunes. The contract for the building of *Black Adder* was signed at the end of June, 1869, and she was launched in March, 1870, having been built to the highest requirements at Lloyd's for iron ships, with all scantlings and materials of the best and a complete East India outfit for a full-rigged ship.

The first evil omen in *Black Adder's* life was the small insignificant fact that her second mate, on leaving his home in Limehouse to join his ship after her launch, found that he had left his purse behind and turned back to get it. And he was greeted at the door by his mother, a sailor's daughter and sailor's wife, with these words—"You should never have turned back. That ship will never be lucky." No prophecy ever proved more true.

They began to load the *Black Adder* before the masts were in her or even a mate appointed, so this same second mate, a boy only just out of his time, had to keep one eye on the cargo tallying and the other on the riggers. The topmasts were hardly on end and the backstays set up before old John Willis came along and noticed that the backstays were slack, upon which he immediately came to the conclusion that they could not have been properly set up, and letting loose some of the language for which he was celebrated, he ordered the second mate to see that the riggers set them up taut. The young officer replied with equal heat that as he was tallying at two hatches at once he could hardly be responsible for the rigger's work. Upon which Willis thundered that he was no use as an officer unless he could superintend three or four things at once. This put the second mate's back up, and clapping on a luff tackle, he soon had those backstays as taut as harpstrings.

On the following morning Captain John again sent for him and pointed to the backstays, which were as slack as ever. The second mate explained what he had done, whereupon the cause of this strange slacking up was looked for and it was soon seen that a great mistake had been made by the builders.

It was found that the cheeks which were bolted to the mast to support the trestle-trees did not extend forward to their full length but only half way, so that a foot of iron angle was left without support, with the result that the fid of the topmast rested on the angle irons outside the cheeks, and, directly the topsails yards were crossed, the trestle-trees bent right down. On the mistake being discovered we may be sure that someone got into hot water with peppery old John Willis, especially as there were no fewer than three captains supposed to be looking after the outfitting of the ship.

The next question was what to do to remedy the defect. Undoubtedly what they should have done was to have sent down the topmasts and bolted on fresh cheeks even if the lower masts had to be lifted out to do it.

What they did do was to bolt on false cheeks and put stays from the ends of the trestle-trees to the caps of the lower masts, but, of course, they could not straighten out the angle irons, and this was ultimately the cause of her dismasting.

She was very heavily rigged with a mainyard 80 feet long and her other yards in proportion up to a main skysail, and this meant a heavy strain on the defective work. Captain Campbell, the rigger, was evidently very distrustful of the makeshift, as he cautioned the two mates with these words, "You are both young men, be very careful."

The unlucky *Black Adder* was not to get away from London without further trouble; indeed, she had a narrow escape of being sunk in dock. When she was nearly loaded the second mate happened to be going round the after hold when he noticed daylight coming in round the flange of a pipe which was only 6 inches above the water. On examination it was found that there was nearly ¼ inch clear space round the flange, and she had to be tripped by the head before this could be put right.

At last the ship sailed, and she was no sooner at sea than she kept her crew constantly busy setting up her topmast and topgallant backstays, until, when they were losing the S.E. trades with the "roaring forties" close ahead, they took the precaution of putting the stream chain under the heel of the main topmast, which was the worst of the three, and over the head of the lower mast. But in spite of this, the first bit of a blow showed that the trouble aloft was very serious

However, *Black Adder* was as unlucky in her captain as in everything else, for though the rivets through the cheeks on the main were all slackening up he would do nothing except send the skysail yard down on deck. The carpenter wanted to bore a hole through the cheeks and put the winch handle through it, but this the senseless skipper would not hear of. The main topgallant mast should also have been sent down.

Black Adder's first blow was not very severe, but it necessitated extra chain lashings aloft. And then a severe gale overtook her, and she had to be laid to the wind under a main lower topsail.

That night, in the first watch, the steering gear went wrong and had to be remedied. Then just before eight bells (midnight), the fool-headed skipper came on deck and found that the wind had fallen away and shifted a few points, upon which he ordered the jib to be set and all hands to be called for wearing ship. This, considering the state of things aloft, was a most foolish and risky manoeuvre, and it should have been evident to the most inexperienced that with no sail on her to steady her she was going to roll badly as soon as she got before the sea. However, the old man was in such a hurry to fall out of the frying pan into the fire that he ordered the helm to be put up before the watch below reached the deck. In squaring the mainyard the young second mate, whose watch it was, and who fully realised the danger, was careful to slack away inch by inch, so as not to give the yard any play, in spite of the fact that his captain was growling at the slowness with which they were squaring the yard. The second mate, however, paid no heed to his captain, but kept his eyes glued on the *Black Adder's* maintop. Then as the wind came aft the expected happened. There was a flash of fire aloft. The chains had parted! In a moment the cheeks fell adrift from the mast, and down fell the lower rigging in a bight. Without its support the mainmast began to heel at a greater angle on every roll, and the ominous words "Stand clear" and "Look out for yourselves" ran along the line of men at the braces.

At this moment the mate appeared on deck, and, taking in the situation at a glance, sang out—

"Get an axe and nick the mast."

"It is nicked right enough," coolly replied the second mate. And sure enough it was. Not being wedged in the partners at the main deck, the mast was buckling below in the 'tween decks; and as it heeled further and further it burst up the main deck, and smashed some cases of glass, which were stowed round it, with a tremendous crashing. This continued for a few minutes, with the ship rolling heavily in the trough of the sea. Then the mast, hanging at an angle of 45 degrees over her port side, seemed to steady her a little, and her crew ventured to get axes to cut away the lanyards; but before they could use them an extra big roller came along and sent the mast right over her starboard side. In its fall it tore up the main deck planking and brought its broken heel up flush with the rail.

With the braces all gone, the after yards—which were now flying first to port and then to starboard every time she rolled—threatened to come clattering down from aloft. Then, as she began to come head on to the sea, the mizen mast itself began to sway ominously fore and aft.

Both watches were now hard at work, and whilst the starboard watch cut the mainmast adrift the port clapped a luff tackle on to the mizen stay in an effort to save that mast. Luckily, for all concerned, the mainmast sank clear of the *Black Adder's* bottom when released from the rigging which was holding it. This fact was made evident by the way in which the main and topsail braces unrove and followed it down into the depths.

And now all hands tailed on to the tackle that had been put on the mizen stay, and as she plunged forward took in the slack; but, unfortunately, the man who should have taken a turn over the pin slipped upon the wet deck, and, as she lifted aft, the two watches could not hold her, so away went the mizen mast. It fell across the taffrail, just missing the wheel, and broke on the rail. Then the rattle of gear and yards over the side began to bump under her quarter, and the rudder began to lift in the most ominous manner. But before this could be attended to the part of the mizen lower mast, which lay across the poop, had to be got rid of, and at the same time the foremast which, having all sail furled and being relieved of the weight of the main braces, upper stays, etc., was sagging forward with its shrouds all slack—had to be saved from following after the main and mizen masts. The mate and his watch at once set about swifting in the fore rigging, whilst the starboard crowd undertook the dangerous work of cutting the mizen adrift, and the carpenter got up some sails and spread them over the hole made when the deck was burst up.

All this time the seas were pouring over all and down into the hold, and this so unnerved the "old man" that he disappeared below and was not seen again until late next day. However, the mate succeeded in saving the foremast, and the second mate in clearing away the wreck of the mizen, though the mast threatened to roll over the second mate and a hand, who had the cutting away of the starboards lanyards, the two just managing to vault clear as it went over the side.

The *Black Adder* was then put before the wind, whilst the two mates retired for a smoke to discuss their next proceedings. Both the fore topsail yards had been broken at the tie by the weight of the main braces when the mainmast went, and the fore topgallant and royal yards were all adrift. However, it was impossible to do any work aloft until daylight. It was also found that the wreck of the mizen had knocked a big hole under her quarter.

As soon as it was light enough to see, the second mate and bo's'n with four hands clambered aloft, the second mate attending to the topsail yards and the bo's'n to the topgallant. This was a most ticklish and difficult business, for the men had all they knew to hang on, for the mast was so insecurely stayed that on one roll the rigging fell in bights to leeward, only to tauten on the return roll with a jerk like that of a gigantic catapult, which did its best to shoot the men off into the sea. A mast rope was got aloft and rove, but they were afraid to hoist the lower yards clear of the stays for fear of bringing the whole mast down, so in order to lower the broken yardarms to the deck they had to cut through the iron jackstays and the sails themselves, At last, at 2 p.m., the topsail yards and topgallant yard were safely landed on deck after a most arduous and dangerous morning's work.

Two men were next sent up to send down the royal yard, a very nasty job, which one of them, Andersen, a Swede, was very loath to tackle; but the other, a native of Deal, named Stevens, called out heartily, "Come along, mate," and up they went. Unfortunately, they forgot to put a guy on the yard before unparelling, and as soon as the yard was freed it flew away from the masthead at a great angle. Below there was a general cry of "Stand from under!" Meanwhile, of the two men aloft, Andersen was clear, but Stevens was still at the topgallant masthead, and he only just had time to slide down the mast and get on the topmast cap when the mast broke 2 feet above him and by a miracle cleared him.

The *Black Adder* was now fairly dismantled, having nothing left aloft but the foreyard. She was 2000 miles from the Cape and 1500 from Rio, so as the wind was southerly and she had a big hole in her quarter her head was turned towards Rio.

Two jury masts were rigged, and a topmast stunsail and a staysail set on each. Then the fore topgallant yard was lashed to the fore topmast, and the skysail yard, which had luckily been sent down on deck before the dismasting, was lashed to the stump of the topgallant mast. And with these three sails forward they were able to set a topmast and lower stunsail.

But by the time the jury rig was in working order—three days, to be exact—the wind shifted and came out of the west, so it was decided to run for Simon's Bay. At this the crew refused duty, saying that the ship was not in a fit condition to go near the Cape. However after 12 hours' rest they thought better of it and turned to again.

On her way to the Cape the *Black Adder* fell in with the *St. Mungo* of Glasgow at daybreak one morning. The *Black Adder* had the wind abeam, and the *St. Mungo*, seeing her jury rig, bore down from to windward and got under her stern with the intention of speaking her, but he could not catch her even with her jury lash up.

Luckily for the lame duck the wind came south-east off the Cape which enabled her to sail up False Bay without help. However, even when her anchor was on the ground, she was not free from mishap.

First of all, in making the anchorage, she fouled a hulk, then on the following morning a barque in getting underweigh collided with her, and before she left Simon's Bay still another vessel ran foul of her.

After some delay new masts and spars were sent out and she was re-rigged, though the new masts and spars did not measure anything like the old ones; however, at last, she proceeded for Shanghai.

Then, when she was half-way up the China Sea, she was run into by the French mail steamer and cut down to the water's edge, but once again she survived her misfortune and crawled into Shanghai.

From Shanghai, on being again repaired, she sailed for Penang to load home. Whilst at Penang she may be said to have been lucky in only losing her jibboom in a collision, and she eventually sailed for London on 23rd July, 1871, arriving home on 17th November, 117 days out.

Meanwhile the underwriters had refused to pay her claim on the ground that she was unseaworthy because the cheeks of her masts had been secured with tap screws instead of rivets, and the underwriters won the day. Whereupon John Willis went for the builders, and lawsuits over the unlucky ship dragged on for 18 months.

Black Adder's incompetent captain was at once discharged on her arrival home and placed by Captain Moore, a very experienced man in the China trade. He took her out to Shanghai, and brought her home from Foochow in 123 days; it was a lucky voyage for the bewitched ship, as she only had one collision in which she lost her mizen topgallant mast.

Moore then left her to take over the *Cutty Sark* and was succeeded by Sam Bissett, who had been mate of her on her first voyage. He took her out to Sydney, and then loaded coal for Shanghai. In crossing the Pacific, *Black Adder* ran into a typhoon and was thrown on her beam ends. Bisset cut away the main and mizen masts, but she would not right herself and it was then found that the coal had shifted. However, after great exertions, the coal was trimmed over and she managed to struggle into Shanghai. After her lost masts and spars had again been replaced with smaller ones, she went to Iloilo and loaded for Boston.

Black Adder left Iloilo on 22nd October, 1873, in company with the *Albyn's Isle*, a barque which was bound to Melbourne. With the N.E. monsoon apparently firmly set in, the two vessels steered to enter the China Sea by the Balabac Strait. Two days out, however, when off the St. Michel's Island, the wind shifted to the westward, and they had to work towards Balabac against a strong S.E. current in light airs, calms and sudden squalls. They did not reach Banguey Island until 2nd November, when both vessels anchored for the night. They got underweigh again at daylight on the 3rd, but could make very little headway against the strong easterly current. The wind came in squalls from S.W. with torrents of rain and thick cloudy weather.

About 4 p.m. *Black Adder* stuck on an uncharted reef—where there should have been 30 fathoms of water—and began to pound heavily on the rocky bottom. Captain Bisset at once began throwing cargo overboard in order to lighten the vessel, but finding this without avail at length transferred his crew to the *Albyn's Isle*, which had run down to be of assistance. But no sooner were *Black Adder's* crew aboard the barque, before a squall off the land took the stranded ship full aback, and backed her off the reef, and away she went as if steered by some demon. It took the *Albyn's Isle* 4 hours to catch her in order to put Captain Bisset and his crew aboard. Luckily owing to the extra strength of her iron plates *Black Adder* sustained no injury from her pounding, but her bottom was very foul and she made a terribly long passage to Boston.

On her fourth voyage, a Captain White tried his hand at her and soon found what he was in for, as she broke her windlass when anchored off the North Foreland and nearly killed her new master besides losing her anchor and chain.

And so she went on, always just escaping destruction in spite of numerous mishaps. After her fatal voyages in the China trade, Willis put her into the Sydney trade, where she was well known for many years. Finally in the nineties, when Willis's fleet were sold. she went to the Norwegians and was lost at Bahia on 9th April, 1905.

"Hallowe'en."

Whilst John Willis disputed in the Law Courts with underwriters and builders, *Black Adder's* sister ship, *Hallowe'en*, lay finished, but not delivered, for it was not until the final lawsuit had been settled that she was handed over. Then, having loaded for Sydney, she set sail in charge of Captain Watt of Peterhead; but, to everyone's surprise, she put back from the Chapman, it being alleged that she was leaking badly and that the pilot, Daddy Daines, had refused to proceed in spite of all the importunities of the furious captain, who scoffed at the very idea of putting back for a little water in the well.

Hallowe'en was re-docked and the cargo taken out, when it was found that while she had lain waiting to be handed over by the builders, an amount of rain water had run into the hold, which, owing to dirt being carelessly left in the limbers, had not been able to flow into the well, and her movement in a seaway had caused this water to free itself so that her well was suddenly discovered to be full of water, which gave the impression, of course, that she had sprung a bad leak.

On making a fresh start, she again roused the anxiety of those interested in her by washing away her head boards in the mouth of the Channel. These were picked up, and as she was not spoken during the passage grave fears were entertained for her safety. However, anxiety was changed to jubilance when the news arrived that she had made Sydney in the wonderful time of 69 days, her abstract showing that—

She sailed on 1st July, 1872; crossed the line in long. 27° W., on 20th July, 19 days out; crossed the meridian of the Cape in lat. 42° S., on 10th August, 40 days out; and arrived Sydney, on 8th September, 24 days from the Cape Meridian. In this passage *Hallowe'en* proved that she had a remarkable turn of speed especially in light winds.

Her speed in light winds, which was most unusual for an iron ship, was chiefly attributed to the way in which her masts were raked. This was in Chinese-junk fashion, the foremast having a slight rake forward, the mainmast being upright, and the mizen raked aft. This gave more spread to her very big sail plan, and kept her sails from blanketing each other, which is one of the drawbacks when all three masts are raked alike.

Hallowe'en, though she so closely resembled *Cutty Sark* in her underwater body, never was able to make such big 24-hour runs as the composite clipper, though she was an exceedingly fast ship all round in anything up to a fresh breeze.

She was also a very dry ship and a good sea boat, and under Captain Watt made passages both in the Australian and China runs which have never been excelled.

"Lothair."

The last out-and-out composite tea clipper of the type of *Ariel* to be built was the beautiful little *Lothair*. She, however, kept mostly to the Japan and Manila trade, and as a rule went to New York instead of London, so that coming on the scene so late in the day and rarely joining in with the crack ships which got the first London teas, she has been rather overlooked when the records of the tea clippers have been spoken about.

However, with regard to her speed, here is the testimony of an American skipper, who was once a well-known passage maker in the American Cape Horn trade—

"The fastest ship, I think, that ever left the ways was the *Lothair*. I'll tell you what happened to me once: I was second mate of a Newbury Port ship and we were running our easting down bound out to Canton, and were somewhere near Tristan d'Acunha, when we sighted a vessel astern. It was blowing hard from the nor'west, and the next time I looked a couple of hours later, there was the ship close on our quarter, and we doing 12 knots. 'Holy jiggers,' says I to the mate, 'there's the *Flying Dutchman*.'

" 'No,' says he 'it's the *Thermopylae*.' But when she was abeam a little later, she hoisted her name, the *Lothair*, and it's been my opinion ever since that she was making close to 17 knots."

Like *Hallowe'en* however, *Lothair* was really at her best in light winds, when even *Taeping* or *Ariel* would have found her a tough nut to crack, but in heavy weather she was not large or powerful enough to equal the records of *Cutty Sark*.

She was very heavily sparred, crossing a main skysail, and under such hard drivers as Captains Orchard and Tom Boulton she made some splendid passages in the New York trade.

Outward Passages in 1870-71.

The best outward passage to China in the winter of 1870-71 was made by the *Cutty Sark*, the following being an abstract from her log—

November 10—3 a.m., passed through the Downs. 2 p.m., signalled St. Catherine's. 9 p.m., Start Point north 8 miles; wind north moderate.

November 29—Crossed the equator in 25° W.; 19 days out.

December 17—Crossed meridian of Greenwich in 41° 42′ S., 37 days out.

December 19—Run 320. Strong northerly breeze.

December 21—Crossed meridian of the Cape in 43° 50′ S.; 41 days out.

December 23—Run 318. Strong S.W. breeze.

December 31—Run 326. Strong northerly breeze.

January 20—5 a.m., sighted Sandalwood Island.

January 24—In Ombay Passage in company with *Titania* and *Taeping*. (*Titania* left London 27th October and *Taeping* 17th October.)

January 28—A moderate breeze right down the Manipa Strait. At 4 a.m. a heavy thunder squall when just in the narrowest part of the Strait. Night pitchy dark and with the darkness and lightning it was impossible to see anything. Hove-to until daylight. At 6 a.m. *Taeping* in sight and American clipper *Surprise*, New York to Shanghai 95 days. Noon, *Titania* came through the Strait with a fair wind whilst *Cutty Sark* lay becalmed outside.

January 29—Baffling airs and calms. *Titania* and *Taeping* in company to eastward. *Surprise* out of sight astern.

January 30—*Titania* and *Taeping* in company close at hand. (Just after noon Captain Moodie went on board *Titania* and returned at 1.30 p.m. with Captain Dowdy, who returned to his ship at 3.30 p.m.)

January 31—Faint airs and calms. *Titania* and *Taeping* a few miles off.

February 1—not a breath of wind up to noon. *Titania*, *Taeping* and *Surprise* in company. P.M. breeze from N.W. *Titania* and *Taeping* dropped out of sight astern.

February 2—Faint airs. *Surprise* still in sight.

February 4—10 a.m., canoes of natives came off from North Island to trade cocoanuts and small shells for old iron.

February 5—First of N.E. monsoon. Unsettled and squally.

February 13—Squally. Fore topsail tie broke, which broke cap on lower masthead.

February 14—Strong gale and head sea. Split main topmast staysail and broke main topsail tie. Blowing hard with snow at times.

February 16—4 p.m., got a pilot; 6.30 p.m., passed the lightvessel; 9.30 p.m., anchored in the river.

98 days out from London to Shanghai.

Cutty Sark's great rival, *Thermopylae*, also did a wonderful passage out, after a great race with *Norman Court*, *Thermopylae*, being bound to Melbourne and *Norman Court* to Shanghai.

It was bitterly cold weather when *Norman Court* towed down the Thames, and on 23rd December she ran into the Downs in a snow blizzard, having carried away her main topgallant yard and narrowly escaped piling up on the Brake Sands. All Christmas Eve she lay in the Downs frost-bound and with a foot of snow over all. The same day *Thermopylae* left London. She also had trouble, losing an anchor off the Nore and not getting through the Downs until the 29th.

Norman Court managed to get away from the Downs on Christmas morning with a northerly wind, but, whilst she was held up by a couple of days' doldrum weather off the Lizard, *Thermopylae* came romping down Channel with a strong nor'-easter behind her, and was nearly up with *Norman Court* before the latter took the same wind.

Norman Court passed Madeira eight days out from the Downs.

On 18th January the equator was crossed by both clippers, *Norman Court* being 19 days and *Thermopylae* 18 days from the Channel.

On 2nd February the two ships were in company in lat. 42° S., long. 13° W., but they parted finally here, as *Thermopylae* ran her easting down on a more southerly parallel. However, though out of sight of each other, the race continued to be of the closest description, and on the day that *Thermopylae* arrived at Melbourne *Norman Court* passed the South Cape, Tasmania. This was on 2nd March, *Thermopylae's* passage being only 65 days.

Meanwhile, *Norman Court* continued to make good running, and, passing Norfolk Island 70 days out, arrived off Sulphur Island, near the Saddle Group, 103 days from the Downs. Here the weather grew so thick that Captain Shewan was obliged to heave to. But about 4 p.m. the fog lifted and showed a ship coming up from the southard with stunsails alow and aloft. This proved to be the *Sir Lancelot*, which had sailed from London a fortnight before *Norman Court*. The two ships were soon close enough to exchange signals, but with nightfall the fog closed down again, and they were obliged to stand off shore together. *Norman Court*, however, tacked at midnight, and, crawling in with the lead, picked up a pilot, and so got into Shanghai a day ahead of *Sir Lancelot*, which performance was considered a great feather in Shewan's cap, and made him the hero of the hour amongst Shanghai shipping people.

Tea Passages of 1871.

Owing to the slump in tea rates, and the increasing competition of steamers using the Suez Canal, the clipper fleet was very much scattered in 1871, only three ships of any racing renown loading at Foochow, whilst the veterans *Fiery Cross* and *Flying Spur* deserted the London trade for that of New York and loaded in Yokohama. *Sir Lancelot*, also, and *Eme* took Shanghai tea to America; and *Belted Will*, which had been faithful to Whampoa for so long, went to Manila for a cargo. Then *Kaisow* went to Amsterdam from Batavia. And other celebrated ships, such as *Leander*, *Wylo*, *Windhover*, *Taeping*, and *Serica*, were all missing from the racing fleet, of which the chief starters were:—

Ship	Captain	Port Left	Left	Passed Anjer	Arrived	Days Out
Thermopylae ..	Kemball	Shanghai	June 22	July 22	Oct. 6	106
Forward Ho ..	Hossack	,,	,, 24	—	,, 20	118
Undine	Scott	,,	,, 27	—	,, 16	111
Titania	Dowdy	Foochow	July 1	July 28	,, 2	93
Maitland	Reid	,,	,, 8	—	Nov. 9	124
Norman Court ..	Shewan	Macao	,, 15	Aug. 8	,, 3	111
Lahloo	Smith	Foochow	,, 27	Sept. 2	,, 15	111
Cutty Sark ..	Moodie	Shanghai	Sept. 4	Oct. 5	Dec. 20	107
Ariel	Talbot	,,	,, 4	—	,, 27	114

Titania was, of course, the heroine of the year, and actually passed *Thermopylae* between Anjer and the Channel, a performance that Captain Dowdy had a just right to be proud of.

Cutty Sark was again rather unlucky with her winds, and experienced very bad weather rounding Agulhas. However, she made up time on the last lap by running from the Western Isles to the Start in 7 days, on one of which, 18th September, she made 323 miles in the 24 hours before a strong S.W. wind and heavy sea.

It was the beautiful *Ariel's* last race, as she was posted as missing when outward bound in 1872, the general belief being that she was badly pooped and broached to when running her easting down. She was always a ticklish vessel to handle, especially when running heavily, owing to her fineness aft, and that she broached to and foundered with all hands seems to be the most likely explanation of her disappearance.

The China Trade in 1872.

The year 1872 was fatal to three other well-known clippers. The old *Ellen Rodger* was wrecked in the Java Seas, *Yangtze* disappeared from the register, and *Lahloo,* through the fault of her second mate, was piled up on Sandalwood Island on 30th July. The year, however, opened well for the China clippers. Freights on the coast were booming, and the rates for rice from Shanghai to Swatow and Whampoa were altogether phenomenal in the early months of the year. In the previous year the clippers had been carrying rice between the Chinese ports at 12 cents a pical, but in January and February, 1872, 80 and 90 cents were freely paid.

Titania, arriving in February after a quick passage out, did so well that she was able to leave for home on 25th May.

Undine, however, was the lucky ship of the year. She secured the record rate of 102 cents a pical from Shanghai to Swatow, and then, getting on the berth early, loaded tea at £4. 10s., and on the whole voyage all but cleared her own value in profit.

Norman Court, arriving just at the end of the high rates, succeeded in getting 40 cents a pical or 32s. 6d. a ton for rice from Shanghai to Swatow. and then went up to Japan and took Japanese rice to Hong Kong at 50 cents before going on to Whampoa to load tea.

Cutty Sark and *Sir Lancelot* were, however, too late to participate in these good rates. The two ships did not leave London until 8th and 10th February respectively, but had a keen race out to Shanghai, *Cutty Sark* managing to beat her redoubtable opponent by just a week and arriving on 28th May. However, it is only fair to state that on her arrival in New York, November of 1871, *Sir Lancelot* had had her racing kentledge (100 tons) taken out to increase her deadweight capacity.

Of the clippers which did not load for London, *Serica* left Hong Kong for Monte Video; *Belted Will* left Iloilo for Boston; *Wylo* Yokohama for New York; *Forward Ho* Manila for New York; *Chinaman* Shanghai for New York; and *Flying Spur* and *Black Prince* left Foochow for New York.

Tea Passage of 1872.

Ship	Captain	Port Left	Left	Passed Anjer	Arrived	Days Out
Titania	Dowdy	Macao	May 25	—	Sept. 19	116
Cutty Sark ..	Moodie	Shanghai	June 18	July 19	Oct. 18	122
Thermopylae ..	Kemball	,,	,, 18	July 19	,, 11	115
Undine	Shearer	,,	,, 24	Aug. 17	,, 17	115
Blackadder ..	Moore	Foochow	,, 27	Aug. 9	,, 28	123
Sir Lancelot ..	Edmonds	,,	July 7	—	Nov. 6	122
Maitland ..	Reid	,,	,, 19	—	,, 6	110
Harlaw	—	,,	Aug. 1	Sept. 8	,, 21	112
Doune Castle ..	Erskine	Shanghai	,, 1	—	Dec. 2	123
Falcon	Dunn	Macao	,, 4	Sept. 6	Nov. 21	109
Taitsing	Bloomfield	Shanghai	,, 8	—	Dec. 2	116
Norman Court ..	Shewan	Macao	Sept. 14	Oct. 5	,, 18	95
Ziba	Green	Foochow	Oct. 1	—	Jan. 16	107
Eme	Sproule	,,	,, 4	Oct. 26	,, 14	102
Fiery Cross ..	Murray	Shanghai	Dec. 5	—	April 2	119

Norman Court distinguished herself this year, not only by making the best passage home, but by weathering out a very severe typhoon in the China Seas.

"Norman Court" in a Typhoon.

As usual, she took her last 250 tons of tea aboard at Macao, after loading the rest at Whampoa, and on 14th September the last lighter came off, but the wind was blowing so hard from the N.E. and the sea was so rough that it was unable to lie alongside. However, it was brought under the *Norman Court's* lee, and the chests tossed aboard and caught by hand.

By 3 p.m. the weather had grown worse with heavy squalls, and the agent, who had come aboard to say goodbye and get the bills of lading signed, had the greatest difficulty in regaining the junk, which took five hours putting him ashore, and a very perilous five hours it was.

Meanwhile, the *Norman Court* was finding it no easy job, to get her anchor with such a strong nor'-easter blowing. However, by 6 p.m. she had catted it, and, setting all sail to her main royal, sped away dead before the wind at a tremendous pace.

Though the weather looked wild the glass had not begun to fall, and no one anticipated what was coming. But as soon as she cleared the islands the *Norman Court* encountered a nasty sea rolling up on her port beam. Old Captain Shewan who was not at all well, still put his faith in the glass, and, leaving the ship in charge of his son, went below to rest, with the usual request to be called if there was any change.

Young Shewan lost no time in preparing for bad weather, which, in spite of the glass, he believed to be coming. The boats were got off in the skids and lashed on chocks to the deck. Extra lashings were put on the spare spars and extra gaskets aloft. The main royal soon had to come in. And by midnight, the weather was looking wilder than ever, and, a still more ominous sign, the barometer had begun to fall.

At eight bells the old man was called, and, after one look round, he turned to the mate and said:—

"We're in for it. Get the sail off her as quick as you can."

Young Shewan began with the foresail (the mainsail had never been loosed), but by the time that was fast the upper topsails were blowing to ribbons.

By four bells sail had been reduced to a main lower topsail, but the wind had increased to such an extent that the mainsail and other sails, in spite of extra gaskets, were blowing adrift. Until 3 a.m. the wind held in the north-east, but it then began to back very rapidly to north and nor'-west. Shewan was obliged to keep his ship dead before it, and with the easterly sea, *Norman Court* began to plunge into it very heavily, and washed two men under the spare spars, hurting them severely. Captain Shewan then told his son to get the main lower topsail—the only rag set—off her. Yet, though all hands were sent to the braces the yards braced by, and the clew line well manned, as soon as the sheet was started the lower topsail gave one shake and was gone.

By daylight the *Norman Court* was running due east under

bare poles with two men at the wheel; she was going like a mad thing and plunging to the foremast, but shipped no water aft. The sea was like a boiling cauldron, leaping high up on both sides of her and falling over both rails at once. The wind was like a thousand furies and the rain fell in solid sheets.

A whole suit of sails was blown out of the gaskets, the forward brace and other blocks being jammed with lumps of torn canvas, and the service of the backstays, etc., was white with the threads of canvas blown on to and tightly wound round them.

It was with the utmost difficulty that the men remained at the wheel, and in order to get aft to see the course steered, the captains and mates were compelled to lie down and crawl along the deck. After one of these journeys, the mate, as he raised himself, was picked up by the wind like a feather and hurled forward until brought up by the poop rail. The second mate, who had come on deck in only a shirt and trousers, had the shirt ripped off his back and whirled away into the skud-filled sky.

The *Norman Court* ran clean round a circle, and about daylight the centre must have been very close aboard by the rapidity with which the wind shifted. But though the sea heaped up in pyramids on each side of her, she was as lively as a lifeboat, and a passenger who was watching the terrifying scene from the top of the main hatch was full of admiration at her behaviour. He happened to be a seamen who had sailed in the *Lord of the Isles*, and he declared that that ship would never have lived through such a sea.

By 6 a.m. the *Norman Court*, which had been steering S.W. when the typhoon began, was steering N.E., and at 8 o'clock she was steering north. The glass now began to rise and Captain Shewan decided to bring her to.

As they brought her to the wind she lay down so far as tc show that, if this manoeuvre had been attempted earlier, she would have either gone right over or they would have to cut away the sticks to save the ship.

When she came head to sea the lookout, who had not been called off the foc's'le head, was washed aft, and, bringing up against the foremast, nearly broke his back.

The typhoon, though it did not last long, tried the ship to her utmost, but she ran beautifully and behaved to the admiration of her crew, emerging from her buffeting in the most triumphant manner, for beyond the loss of her sails which were blown out of double gaskets, she did not strand a rope yarn.

Her crew, however, did not escape so easily, many of the men being badly knocked about, whilst the two mates were so hoarse from shouting commands in the screaming wind that they could not speak above a whisper for days afterwards.

Having survived this strenuous opening to her passage, *Norman Court* now proceeded to make a splendid run home, as the following abstract will show:—

Norman Court, Macao to London, 1872.

Sept. 14	—Left Macao 6 p.m.	
Sept. 21	—Off Pulo Cambii, Cochin China	7 days out
Sept. 26	—Great Natuna, S. end W. by N. 20 miles	12 ,,
Oct. 2	—Gaspar Island, W by S. 10 miles	18 ,,
Oct. 5	—Passed Anjer 4 p.m.	21 ,,
Oct. 24	—Passed meridian off Cape St. Mary, Madagascar	40 ,,
Nov. 5	—Rounded Cape Agulhas	52 ,,
Nov. 14	—Sighted St. Helena	61 ,,
Nov. 19	—Sighted Ascension	66 ,,
Nov. 23	—Crossed the line 10 p.m.	70 ,,
Dec. 3	—Passed latitude of S. Antonio	80 ,,
Dec. 11	—Flores, S. by E. ½ E. Corvo East.. ..	88 ,,
Dec. 17	—Made the Lizard	94 ,,
Dec. 18	—Anchored in the Downs	95 ,,

And the Baring clipper had a magnificent run up Channel. From the Lizard to Dungeness she was only 19 hours. At 3.30 p.m. on 17th December she was off the Lizard and at 11 a.m. next day she picked up a pilot at the Ness.

Altogether *Norman Court* had had a very successful and profitable voyage, and so pleased were Baring Bros. with their vessel that they commissioned Dutton to execute a picture of her at a cost of £100. He chose the moment at which she picked up her pilot off the Ness, and the illustration given here is from a litho of this picture.

Norman Court's typhoon reminds me of the one which caught *Titania* and *Lord Macaulay* inside the Paracels. Captain Care of the *Lord Macaulay* hove to with a specially made sail lashed to the mizen rigging, and she lay so far over that a man could have walked on her side.

Titania ran it out like *Norman Court*, and with a bow wave towering above her rail came foaming by the *Lord Macaulay* like a whale-boat in tow of a whale; indeed, she was so close to the latter that she washed her fore and aft, with the white bone she had in her teeth. Both ships lost their royal masts, and when the two captains met again in London, Captain Care hardly knew his brother skipper, for the latter's hair which, when he left China, had been coal black, was snow white.

The Race Between "Cutty Sark" and "Thermopylae."

A great duel was arranged in 1872 between *Cutty Sark* and *Thermopylae.* Both vessels left Shanghai on the same day and within an hour or two of each other. They were, however, some time in getting clear away owing to fresh gales and thick fogs in which it was impossible to proceed, and *Cutty Sark* did not drop her pilot until 21st June.

They were then held up by calms and fogs until 2 a.m. on the 23rd, when the N.E. monsoon began to blow strong and soon freshened to a gale, which split the *Cutty Sark's* fore topgallant sail to pieces.

The monsoon held until the 26th when at 1 p.m., in lat. 20° 27′ N., long. 114° 43′ E., the two racers were in sight of each other, *Cutty Sark* being in the lead.

On the 28th June they were again together, this time with *Thermopylae* 6 miles to windward of her opponent, the wind being fresh from the S.W. with heavy squalls, but they did not meet again until approaching Gaspar Straits. The weather continued boisterous until the 1st July, up to which date *Cutty Sark* had only had one observation since leaving port.

On the Cochin China Coast the usual land and sea breezes were worked, but crossing to the Natunas fresh gales and squalls and split sails were the experience of both clippers.

On 15th July in 108° 18′ E. on the equator, *Thermopylae* sighted *Cutty Sark* about 8 miles ahead, but gradually fell astern, and on the following morning *Cutty Sark* could only just be seen from the fore topsail yard bearing S.E. At 10 a.m. on 17th July *Cutty Sark* led *Thermopylae* through Stolzes Channel, but on the 18th some unfriendly waterspouts compelled the former to bear up out of her course and take in sail and this let *Thermopylae* up. At 6 a.m. on the 19th both ships arrived off Anjer, *Thermopylae* now having a lead of 1½ miles. Here *Cutty Sark* was hove to for a couple of hours whilst Captain Moodie went ashore with letters.

At noon on the 20th *Thermopylae* was 3 miles W. by S. of *Cutty Sark*, both vessels being hung up by calms and baffling airs. And it was not until the 26th, with Keeling Cocos Island in sight to the nor'rard that there was any strength in the S.E. trade; from this point, however, the wind came fresh from the E.S.E. and stunsail booms began to crack like carrots.

This was the sort of weather that *Cutty Sark* revelled in, and she went flying to the front with three consecutive runs of 340, 327 and 320 miles. She carried the trades until 7th August, when at 1 p.m. the wind suddenly took off as if cut by a knife, and remained calm and baffling until the 9th when it commenced to breeze up rapidly from the S.W.

The 11th August found *Cutty Sark* battling with a strong westerly gale, but with a good lead of *Thermopylae*. From this date, however, the weather fought for the latter, and the following quotations from Captain Moodie's private log will show the bad luck which attended *Cutty Sark* in losing her rudder.

August 13—Lat. 34° 3′ S., long. 28° 7′ E. Distance 83 miles. Strong gale from N.E. At 5 a.m. the wind hauled to west. Rest of day blowing a very heavy gale. Fore and main lower topsails went to pieces.

August 14—Lat. 34° 6′ S., long. 28° 7′ E. Heavy gale from W. with severe squalls and tremendous sea.

August 15—Lat. 34° 26′ S., long. 28° 1′ E. At 6.30 a.m. a heavy sea struck the rudder and carried it away from the trunk downwards. Noon, wind more moderate, tried a spar over the stern but would not steer the ship. Thereupon began construction of a jury rudder with a spare spar 70 feet long.*

---

* Captain Moodie's description of his jury rudder.—"The making of the rudder was, however, only the simple part of it, the connecting it to the post and securing it to the ship so that it would work and be of sufficient strength for use when placed was the most difficult part of the job. The connection was made by putting eye-bolts in both rudder post and rudder, and placing them so that the one would just clear the other; a large bolt (an awning stanchion)

August 16—34° 13′ S., 28° 24′ E. Light winds from south. P.M., strong breeze from E.N.E. Constructing jury rudder and sternpost as fast as possible.

August 17—34° 43′ S., 28° 25′ E. Strong winds from east to E.S.E. Constructing jury rudder and sternpost.

August 18—34° 58′ S., 28° 11′ E. Strong winds from E.N.E. Constructing jury rudder and sternpost.

August 19—34° 51′ S., 27° 58′ E. Strong winds from N.E. Constructing jury rudder and sternpost.

August 20—34° 38′ S., 27° 36′ E. Light wind from westward. Noon, strong westerly breeze and clear. About 2 p.m. shipped jury rudder and sternpost, a difficult job as there was a good deal of sea on. (It will be noticed that whilst *Cutty Sark* lay hove to, with her crew working night and day on the jury rudder, fine fair winds, which carried *Thermopylae* round the Cape, were blowing, but no sooner was the rudder ready for shipping into place than the wind chopped round into the west and began to blow up for a further series of head gales.)

August 21—34° 19′ S., 26° 58′ E. Distance 36 miles. Strong westerly gale.

August 23—35° 49′ S., 20° 58′ E. Distance 194 miles. Stiff breeze from south to E.N.E. and sharp head sea. Midnight, wind hauled to N.W. Rounded Cape Agulhas. (On this day *Thermopylae* was in 31° 43′ S., 13° E., 490 miles ahead).

Cutty Sark next had a succession of heavy head gales, which did not let up until the 31st, and sorely tested the capabilities of the jury rudder. The awning stanchions which connected the steering chains to the back of the rudder were carried away, and several of the eye-bolts which held the rudder to the post were broken, but they managed to steer with two wire rope pennants shackled to an eye-bolt placed in the back of the rudder in case of accident to the chains.

---

was then passed through them and clenched on both ends; in this way we had five eye-bolts in each, locked with two strong bolts which would bear a considerable weight. The securing of the whole to the ship was of the next importance, and it was soon apparent that this could not be done in the way usually recommended, viz, by placing chains along the ship's bottom and leading into the hawse pipes; in the first place, the *Cutty Sark* is too sharp for chain to lie along the keel, and in the next place the length of the ship is too great; it would be difficult to bind the post tightly to the vessel owing to the great length of chain. I therefore concluded to take both the guys to the after mooring pipe, fitting the lower one with a bridle under the keel, 16 feet from the heel of the ship, so that from the post to the bridle there was a little down-pull which prevented post and rudder from rising. The next thing to be done was to get the steering gear secure to the rudder, for the trunk was too small to admit anything but the false sternpost, which came about 2½ feet above the deck, and being wedged round formed a good support. The steering gear had therefore to be secured to the back of the rudder and led to a spar placed across the ship, about 15 feet before the taffrail, which led the steering chains clear of the counter, and then inboard to the wheel. Of course, all the gear was attached to both rudder and post before they were put over the stern. Having a small model of the ship I took all the measurements for the chains by that, which enabled me to place them pretty near the truth.

The jury rudder however carried *Cutty Sark* to 7° 28′ N., 20° 37′ W., without further accident. The ship was found to steer very well with the wind right aft, but with strong beam winds and when going anything over 10 knots the rudder was not nearly so efficient, and it was often necessary to keep the ship down to about 8 knots.

On 1st September in 30° 44, S., 12° 24, E., the succession of fierce northerly gales at last grew tired of buffeting the lame duck and the normal weather for running down to St. Helena set in. The island was passed at 9 a.m. on 9th September, and on the 15th *Cutty Sark* crossed the line. Her best runs between 1st September and this date were 210, 211, 214, 226, 227, 221 and 207, pretty good work for a ship which was not allowed to do more than 8 knots.

All this time, however, the jury rudder was gradually breaking its fastenings and on 20th September the last of the eye-bolts holding the rudder to the post gave way and the whole contrivance had to be hoisted up for repairs. Captain Moodie was now so short of material that he had to shape flat pieces of iron so that they would work on the iron stanchions instead of the eye-bolts. The repairs were smartly done and on the following day the jury rudder was once more ready for lowering.

On the first occasion a kedge anchor of 5½ cwt. had been used to sink it into place, but owing to the bad sea running this had been lost. On 21st September Captain Moodie determined to fix the post and rudder in place without using any weight to sink it. When all was ready the sails were filled and the ship given a little headway, the rudder and post were then lowered and streamed right astern, the rudder was then hauled close to the trunk and the sails laid aback. As the ship lost headway, the weight of the chains partially sank the rudder, then as the ship slowly gathered sternway and the slack of the guys were hauled in, the heel of the rudder sank and allowed the head to be easily hauled up through the trunk. This operation is very easy to write about, but in its proper execution it required such seamanship as is hardly known nowadays.

Cutty Sark had fine strong N.E. trades to within a day of the Western Isles, but, unfortunately, had to be kept down to a speed of 300 miles a day, as beyond that her jury rudder could not control her.

On the last lap of the passage she unfortunately met with strong winds and gales from the nor'rard and eastward, and on 12th October, the day that *Thermopylae* arrived in the Downs, she was battling against a fresh N.N.E. gale in 45° 37′ N., 13° 26′ W. This gale lasted until *Cutty Sark* also reached the Downs on 18th October, less than a week behind her rival, for which fine performance Captain Moodie received great praise in shipping circles. Indeed though *Thermopylae* arrived first, all the honours of the race belonged to *Cutty Sark* for she was hove to for more than 6 days whilst the jury rudder was being made. And between the day on which she lost her rudder and that of her arrival, she wasted 11 days making 139 miles, added to which, when she had a chance to go ahead, her speed had to be reduced to 8 knots or half what she was capable of doing. It therefore seemed pretty certain that, but for her accident, *Cutty Sark* must have beaten *Thermopylae* by several days. This race between *Cutty Sark* and *Thermopylae* was the beginning of a life-long rivalry in which it is difficult to say which came out on top, as whilst during the seventies *Thermopylae* made the best passages, during the eighties *Cutty Sark* made the fastest voyages of the two.

Unfortunately for Willis's clipper, Captain Moodie left her on her arrival in 1872, from which date until 1885 her skippers did not bear the reputation of being "sail carriers." In 1885, however, she was taken over by an old sea dog of the Bully Forbes' type, and immediately she began to make wonderful records in the Australian trade. Captain Moodie was succeeded by Captain Moore, late of the *White Adder*, a well-known London man, who had been mate of the *Lammermuir* in her famous race with *Cairngorm*.

In November, 1872, *Cutty Sark* for the first time was laid on the berth to load for the Colonies; in fact, she loaded for Melbourne almost alongside her rival *Thermopylae*. Unfortunately, she was not quite ready in time to get away with *Thermopylae*, nevertheless the two ships made a very close race of it, their times being—

Thermopylae left London November 14; dropped pilot off Dartmouth November 17; arrived Melbourne January 27, 1873—71 days from pilot.

Cutty Sark left London November 25; dropped pilot off Dartmouth November 28: arrived Melbourne February 11, 1873—75 days from pilot.

Tea Trade of 1873.

The year 1872 may be said to have been the last year in which there was any real racing amongst the clippers. Henceforward though the captains still did their best to make fast passages, they no longer had any chance of bringing the first teas to market, which were all taken by the racing steamers through the Suez Canal, there was therefore no need for daring feats of navigation or sail carrying, added to which, as freights fell before the onslaught of the steamers, the clippers grew more and more scattered and only two or three of the most celebrated of them, such as *Thermopylae* and *Cutty Sark* continued to load home in June and July. Many of the others preferred to load later in the year in November and December, when they had the fair monsoon and reduced insurance rates. Others deserted the English market for that of America. Of these the most regular traders to America were *Wylo*, *Kaisow* and *Lothair*, all of whom made passages from China to New York in under 100 days. In 1873, *Leander*, *Wylo*, *Eme*, *Black Prince*, *White Adder*, *Chinaman* and *Falcon* all went to New York. In the United States the evil practice of running the crews out of arriving ships flourished, and I am told that carrying tea to New York was worth £100 to an unscrupulous captain, who would stoop to such methods. Naturally decent captains strongly objected to this slim Yankee device and were not willing either to run out their crews or pay them off in order to put the money straight into the maws of that worst of all land sharks, the New York waterside wolf, notwithstanding the fact that one-third of the spoils were offered as blood money.

Every obstacle was, of course, put in the path of these straight-going skippers. A case in point was that of Captain Brown of *Wylo*, a "white man" and a gentleman, beloved by his crews. On his stubbornly refusing to pay off his men, the result of his honesty was soon apparent. When his ship was loaded and ready to sail he found that he could not get his clearance. In vain he appealed to the British Consul, that representative of the British Empire was helpless, and after being detained for 10 days, Captain Brown was at last obliged to leave unsatisfied.

On his arrival in London he wrote an indignant letter to the *Shipping Gazette*, stating in bitter language that the British flag was of no use to Britishers in New York. The same sort of experience happened to Captain Shewan, senior, and others.

Besides America, Australia was becoming a growing consumer of tea, and through the seventies some fine little clipper barques worthily upheld the racing traditions of the trade by their smart passages from the tea ports to Melbourne and Sydney. Perhaps the best known of these clipper barques were the *William Manson*, built by Duthie of Aberdeen in 1872 for Frazer of Sydney, and the *Mary Blair* built by Duthie in 1870 for Hobart owners. The *William Manson* only registered 366 tons and the *Mary Blair* 311 tons, but these vessels crossed two skysail yards and were sailed for all they were worth. One of their skippers was an especially well-known character, he was a little man of the type of Captain Kettle. On one occasion he hung on to his skysails so long that his men refused to go aloft and furl them, fearing that the masts might go any moment. Thereupon he went up himself and put the gaskets on the sails and on his return to the deck administered a severe thrashing to the men who had refused to tackle the job, as a pointer to the statement "that he gave no man a task which he feared to do himself."

This year the famous *Taeping* came to an end of her career, being wrecked on Ladds Reef.

The following were the chief passages made in 1873:—

Ship	Captain	Port Left	Left	Passed Anjer	Arrived	Days Out
Sir Lancelot ..	Edmonds	Shanghai	June 29	—	Nov. 3	127
Maitland	Reid	Foochow	July 6	—	,, 3	120
Cutty Sark ..	Moore	Shanghai	,, 9	Aug. 20	,, 3	117
Thermopylae ..	Kemball	Shanghai	,, 11	,, 8	Oct. 20	101
Undine	Vowell	Foochow	,, 16	—	Nov. 26	133
Forward Ho ..	Wade	Foochow	,, 23	—	Dec. 1	131
Titania	Hunt	Shanghai	Aug. 2	Sept. 17	,, 18	138
Norman Court ..	Shewan	Foochow	,, 4	,, 5	Nov. 28	116
Kaisow	Anderson	Shanghai	,, 9	—	Dec. 2	115
Hallowe'en	Watt	Shanghai	Nov. 19	Dec. 6	Feb. 16	89
Lothair	Orchard	Whampoa	Dec. 11	,, 16	Mar. 19	98

It will be noticed that *Cutty Sark* and *Thermopylae* again managed to load together, but in the race down the China Sea Captain

Kemball made the good time of 28 days to Anjer, whereas Captain Moore made the very bad one of 42 days.

It is on this part of the passage that the element of luck comes in, besides when the time to Anjer depended more on the captain than the clipper herself. From Anjer home both ships made very fair passages, *Thermopylae* just having the best of it by two days.

Sir Lancelot, without her racing ballast, was not driven as in Robinson's day, and so her time was poor. But *Titania* was the unlucky ship of the year, being no less than 46 days to Anjer and 50 days between St. Helena and the Downs. The abstract log of *Hallowe'en's* magnificent passage will be found in the appendix. She had come across from Sydney to Shanghai in 31 days, which was almost as good as *Thermopylae's* record of 28 days from Newcastle, N.S.W.

Leaving Sydney on 9th August, *Hallowe'en* crossed the line on the 21st, only 12 days out; on 6th September she made a run of 312 miles with the wind fresh at east and all sails set, and at noon on the 9th September she took her pilot off Leaconna. In this passage she showed extraordinary all round speed in light winds. In the passage home, she was lucky in her winds and had very few days with the yards on the backstays, but, nevertheless, her performance was no fluke, as she repeated it within a day or two in her next two voyages.

Best Passages, 1874-1878: Shanghai, Foochow, and Whampoa to London.

1874.

In S.W. monsoon—*Thermopylae* 101 days, *Norman Court* 114, *Cutty Sark* 118.

In N.E. monsoon—*Hallowe'en* 91 days, *Undine* 113.

1875

In S.W. monsoon—*Thermopylae* 115, days, *Cutty Sark* 125, *Sir Lancelot* 125.

In N.E. monsoon—*Hallowe'en* 92 days, *Titania* 100, *Jerusalem* 101.

1876.

In S.W. monsoon—*Cutty Sark* 109 days, *Thermopylae* 119.

In N.E. monsoon—*Hallowe'en* 102 days, *Norman Court* 106.

1877.

In S.W. monsoon—*Thermopylae* 104 days, *Windhover* 121, *Cutty Sark* 127.

In N.E. monsoon—*Jerusalem* 106 days, *Wylo* 111.

1878.

In N.E. monsoon—*Titania* 102 days, *Thermopylae* 110, *Taitsing* 117.

In 1873 tea freights had dropped to such an extent and the eastern trade was so bad that even *Thermopylae* had great difficulty in filling her hold at 30/- per 50 cubic feet. Indeed, *Thermopylae*, *Cutty Sark*, *Hallowe'en*, and other noted ships were all reduced to making trips backwards and forwards between China and Australia, so difficult was it to get a cargo home.

In 1881 *Thermopylae* made her last passage in the tea trade —leaving Foochow on 30th October she arrived in the Downs 107 days out. The same year *Hallowe'en* made a passage of 103 days from Shanghai. And these are the last records worth noting, for by this date all the tea ships which were still afloat had had their wings clipped and crews reduced, with economy as their guiding star and not speed.

But it is always interesting to follow a well-known ship to the end of her days, so I shall now attempt to trace the after life of these beautiful tea clippers before bringing my pen to a halt.

The After-Life of the Tea Clippers.

The *Falcon* deserted the tea trade long before the eclipse, and spent some years sweltering on the West Coast of South America under charter to the Chilean Government; and when she came home to be reclassed in 1873, it was found that her stay on that coast had done her no good. Her keelson proved to be as soft as a cabbage, and the dry rot had to be literally dug out of her hold. It was soon seen that the day of the famous old flyer was over, and with clipped wings and no yards on her mizen she slowly sank into obscurity.

In 1887, however, she was still afloat, owned by the Austrians and disguised under the name of *Sophia Brailli*.

The *Fiery Cross* was sold to the Norwegians about the end of the seventies and was still afloat until comparatively recent years.

Flying Spur, after having her spars reduced drifted ashore on the Martin Vaz Rocks in the early eighties. *Forward Ho* was wrecked in 1881, whilst *Chinaman* was run down and sunk by a steamer in the Yangtze River, in 1880. *Serica* was lost on the Paracels in the spring of 1873, and *Taitsing* was wrecked on the Zanzibar Coast in 1883.

Min was sold to the Hawaiians, and sailed in the island trade, for many years under the name of *W. B. Godfrey*. *Black Prince* was lost in the Java Sea soon after Baring Bros. sold her.

Sir Lancelot survived her sister ship *Ariel* by many years. Her spars were cut down in 1874, 8 feet being taken off her lower masts. Nevertheless she made her next voyage out to China, back to New York and home in nine months and two days. But in January, 1877, she had the further indignity put upon her of having the spars stripped off her mizen mast. Yet even this could not stop her, and on 28th December, 1877, she left Shanghai under Captain Andrew Hepburn, passed Anjer on 15th January, 1878, and arrived at New York on 2nd April, only 95 days out.

On her next voyage she went out to Japan, commanded by Captain Brockenshaw, and it was on the passage out to Yokohama that she picked up the survivors of the Victorian Expedition to New Guinea.

In 1879, under the same commander, she loaded to New Zealand then crossed to China and loaded home from Foochow, arriving in the Thames on 27th February, 1880, 128 days out. This was her last tea passage.

In 1881-2 Captain Shortlands took her out to Honolulu, then across to Astoria, and home round the Horn.

The next few years, from 1882 to 1885, Captain Murdoch Macdonald had her and kept her on charter in the Indian coasting trade. Finally, in 1886, Messrs. MacCunn sold her to Visram Ibrahim of Bombay.

Henceforward she carried a coloured crew and became a "country trader," running chiefly between Bombay, Calcutta, and Mauritius. She was, however, lucky in being commanded by Captain Brebner, whose handbook for the Indian Ocean is well known to seamen trading to the East. He kept her in such beautiful condition that she was known for years as the "Yacht of the Indian Ocean," and during the last phase of her career she received as much admiration as she had ever had in her glorious past. Amongst her greatest admirers were the Admiral in command of the East Indian Squadron (who could never pass her without praising her beauty and gracefulness), Lord Harris, the Governor of Bombay, and the Governor of Mauritius, each of whom paid her the honour of a visit of ceremony with full staff.

Her passages under Captain Brebner were always very good, and as late as the nineties she weathered out no less than four fierce cyclones without sustaining any material damage, which speaks volumes both for Captain Brebner's seamanship and the old clipper's seaworthiness.

The first of these cyclones began on 6th June, 1892, in 7° N., 92° E., the wind going round to the south from S.W. in terrific squalls until the 8th, when *Sir Lancelot* found herself in 11° N., 89° E.

The second cyclone was encountered on the 26th of the same month and year. It overtook *Sir Lancelot* in 8° S., 85° E., and held her in its clutches until the 30th, during which time the wind shifted from south through to east with the usual violent squalls and mountainous sea.

The following is Captain Brebner's account of *Sir Lancelot's* third cyclone:—

"I sailed from Bombay on the 21st October, 1893, and experienced fine weather down the Malabar Coast. On entering the N.W. monsoon regions, it became squally with incessant showers of rain for days, and, on the morning of the 1st November, *Sir Lancelot* ran into a cyclone right-hand semicircle in lat. 9° 10, S., long. 72° 20′ E. The wind was steady at N.W. during the night with hard squalls and very heavy rain. At midnight, I reduced sail to topsails and foresail. At 5 a.m. the wind shifted from N.W. to N., and at 6 o'clock it was N.E. with mountainous seas. I immediately lay to under the lower topsails. At 7 the wind rapidly veered to east, S.E., south and S.W. where it remained steady, and blew a hurricane of much violence. Sails were blown from the yards, the leeside was under water up to the hatches, the bulwarks were washed away nearly the whole length of the ship, the wheel broken into matchwood, skylight stove in and cabin flooded. The squalls were terrific. At 10 p.m. it showed signs of abating. By midnight the storm had passed and the wind shifted to N.W., when storm sails were set to keep the vessel head on to the sea."

Captain Brebner's account of *Sir Lancelot's* fourth cyclone is equally interesting:—

"*Sir Lancelot* left Calcutta on 20th December, 1894, for Mauritius. On the 12th January, 1895, whilst running with fresh S.E. trade and approaching Rodriguez to sight it, the sun and moon were surrounded by halos, and this phenomenon continued till the night of the 13th, the position then being 50 miles north of Rodriguez. It grew squally with heavy rain after midnight. On the morning of the 14th Mauritius was W. by S., 180 miles, the weather then became very thick. I knew that a cyclone lay in the locality. Running to anchor at the Bell Buoy, a dangerous anchorage which necessitated putting to sea should the cyclone strike the island, was not considered advisable. I therefore decided to lay to, set my cyclone compass and watch the wind and barometer. It continued to rain throughout the day and night, the wind being steady from S.E., moderate in force, and barometers steady. On the morning of the 15th conditions were the same. At 11 o'clock the sun shone out brightly and continued so for half-an-hour. I then anticipated some improvement but would not act until the barometer indicated a change for good or bad. At noon things were anything but promising.

"It now became evident that I was in front of an advancing revolving storm. The barometer began to fall rapidly, mountainous seas rolling up from about N.E. with increasing S.E. wind. I then set the various parts of my cyclone compass and saw that I was on the south-west margin of the storm and also in the dangerous quadrant.

"I considered it was now time to act quickly and seriously. Having a fast ship, I decided to take my chance and run across the front of the advancing storm into the navigating quadrant. The two lower topsails were then set and the ship headed N.W., *Sir Lancelot* making 9 knots by patent log and perhaps 11 over the ground. Before the helm was put up two oil bags were placed over each bow and the same over each quarter and she ran comfortably, although the sea was dreadful to behold. At 4 p.m. the wind showed signs of shifting and the barometer was still going down. At sunset, it veered from S.E. to a little west of south. *Sir Lancelot* was shipping much water amidships, but no damage was done, the oil bags working faithfully and being replenished when necessary. I took in the fore lower topsail before it became dark and made up my mind to sacrifice the main one.

"At 8 o'clock the wind was S.S.W., at 10 o'clock S.W., the maintopsail then blew to ribbons. *Sir Lancelot* ran under bare poles until midnight, when the wind veered to west. At 2 a.m. on the 16th it was W.N.W., when the barometer stopped falling, being at 29·3°.

"Between 2 and 5 a.m. it blew a terrific gale, and *Sir Lancelot* took large quantities of water over the stern as she was then on the wrong tack for bowing the sea. To avoid sustaining any damage and to assist the four oil bags, I placed another larger oil bag in a rattan ballast basket, attached the deep sea lead-line to it and ran it out the full length. The basket, which streamed away to windward, served the purpose, as the sea broke lightly afterwards.

"At 6 o'clock the wind veered to N.W., the barometer rose, wind and sea went down, and the weather became finer. The *Sir Lancelot* escaped with a good shaking up and the loss of the main topsail only."

The game little clipper, after having thus defied four cyclones, was destined to be conquered by her fifth. Under Captain Brebner she would no doubt have again vanquished the elements, but, unfortunately, this time she was not in such capable hands.

In April, 1895, she was sold to Persian owners, and left Muscat in September of the same year, commanded by an Arab and deep loaded with salt for Calcutta. As she never arrived at her destination her fate would have no doubt remained a mystery but for the following letter:—

I was the branch pilot in command of the brig *Fame* at the Sandheads, mouth of the Hooghly, on the 1st October, 1895, when we had a very heavy cyclone. The *Sir Lancelot* came up under my lee and asked for a pilot (squalls were coming up heavier and faster), but there was too much sea to send my boat, so I told the captain to get to the southward as soon as he could. She looked to be very deep, with salt from the Red Sea and was making bad weather of it. That afternoon I was on my beam ends, topgallant masts sent down, and there I lay for five hours, double gaskets on all sails and preventer braces. I think the use of oil bags saved my vessel. About 10th October four lascars were picked up dead in the Bay, supposed to be from the *Sir Lancelot*, but she certainly foundered not many miles from me. I was her pilot and sailed her up the Hooghly to Calcutta some ten years prior to this, so was interested in her. W. F. Wawn.

So passed one of the most famous and beautiful of all the tea clippers.

If *Sir Lancelot* had to contend with cyclones in her old age, *Titania* had the severe test of Cape Horn, and the way in which these beautiful creations survived the ordeal of storm and tempest when over 20 years old says much for the perfection of their build.

Titania was bought by the Hudson Bay Company in the early eighties, and under Captain Dandy Dunn voyaged year after year round the Horn to Vancouver and back. Her big sail area was, of course, cut down, yet, even so, with her fine ends she must have been a ticklish vessel to handle amongst the Cape Horn grey beards. However, in spite of ten years' trading round the dreaded Cape Stiff, she survived *Sir Lancelot* and all her contemporaries except *Cutty Sark*, *Thermopylae*, and *Lothair* by many years.

In the early nineties she was bought by a Mrs. Maresca of Castelmare, and henceforth became a familiar object in the ports of Marseilles and South America.

Her debut under the Italian flag was rather unfortunate. She was sold in Australia, and on her way home collided with the s.s. *Courowarra* off Green Cape on 15th April, 1894, her second officer being found to blame. After which for the next 15 years she was to be seen in Naples, Marseilles, or Rio, looking as spick and span as ever, the only change visible to those who had known her in her prime being the reduced spars. She was finally broken up at Marseilles in March, 1910.

Of other notable ships in the tea fleet, *Leander* was still under the British flag in the nineties, being owned by R. Anderson of London. She also traded out East, and after being damaged by two cyclones, in March and April of 1892, she was sold to Muscat Arabs, and was lost about the same time and in the same way as *Sir Lancelot*, foundering with her Arab crew in a cyclone when bound from Muscat to Calcutta with salt.

Undine had a terrible experience in 1882. She was swept bare by an abnormal wave, which took the second mate and his whole watch overboard—Captain Bristow, who had been in her from his apprenticeship, being found dead under her spare spars on the following morning.

Shortly after this tragedy *Undine* was bought by M. Ivetta of Ragusa, and so disappeared into oblivion. *Windhover* remained under the British flag to the end, trading mostly to Australia. On 15th October, 1887, she arrived in San Francisco 44 days from Newcastle, N.S.W., a passage which was within 4 days of the record, and this in spite of reduced canvas and no yards on her mizen mast.

Shortly after this performance, however, she was wrecked on the Australian Coast.

Kaisow also remained under the British flag to the end. On the 14th November, 1890, she left Valparaiso barque-rigged and loaded with manganese ore for the United Kingdom. At 2 a.m. on the 15th she was running under topsails and foresail, when she was struck by a heavy sea which hove her on her beam ends and caused her cargo to shift. She was then 60 miles W.S.W. of Valparaiso. She only just gave her crew time to get clear in one of her lifeboats before she filled and sank, the lifeboat safely making the land on the following day a few miles south of the River Lamari.

Norman Court continued in the tea trade until 1880, though reduced and converted into a barque in 1878.

At the end of 1873, after a passage of 116 days from Foochow, Captain Andrew Shewan gave up the command of *Norman Court* owing to ill-health, and his son took over the Baring clipper. On the passage out to Australia in 1874, young Shewan was able to try his ship against that flyer in light airs, *Kaisow*. The two vessels met off Beachy Head on the 6th January, and were constantly in company right down to the roaring forties, when they were parted by thick weather in 40° S., 2° W., and did not meet again.

On his first passage home in command, fate gave young Shewan a still greater antagonist, namely *Sir Lancelot*. The latter sailed from Shanghai on 18th July, and *Norman Court* from Foochow on 27th July. On 29th July the two ships met in the Formosa Channel, and were in company until 7th August, when they were parted by the tail end of a typhoon. On 25th August they again met off the Borneo Coast and again lost sight of each other.

Norman Court passed Anjer at 8 p.m. on the 30th, and *Sir Lancelot* early on the 31st. It had been the usual wearisome work of squalls and calms in the China Sea, but both vessels made up for lost time in the Indian Ocean, *Norman Court's* run across the trades being especially good, the following being her best week's work:—

Date	Lat.	Long.	Course	Dist.	Winds
Sept. 1	9° 0′ S.	99° 55′ E.	S. 65° W.	308	Strong trades, squalls and showers
„ 2	10 31	95 2	S. 72 W.	303	Strong trades, S.E. to S.S.E., occasional lulls.
„ 3	11 53	89 48	S. 75 W.	319	Strong trades, heavy beam sea.
„ 4	13 21	85 7	S. 73 W.	289	A.M., declining. P.M., fresh. Heavy S.E. swell.
„ 5	14 46	80 29	S. 71½ W.	283	Fresh variable trades, S.S.E. to S.E. by E. Heavy rain.
„ 6	16 18	76 5	S. 70 W.	271	A.M., clearing. P.M., fresh trades.
„ 7	17 56	71 42	S. 69 W.	273	A.M. brisk. P.M. decreasing trades.

This totalled 2046 miles for the week. *Norman Court* was going with topgallant stunsails set most of the time, but it was generally a job to get your soup. For five days she averaged over 300 miles a day, but she had to be watched. Captain Shewan hove the log once when she was running, wind quarterly, with main royal set. He brought her a point or two to windward of her course and let her have the weight of the wind, and she proved to be going 15 knots, which was about the utmost to be got out of her.

On the 25th September *Norman Court* was off Agulhas, a strong N.W. gale blowing, and a great many vessels in company. She crossed the equator on 15th October, 80 days out. The N.E. trades were lost in 23° N., 31° W., on 25th October, and she had the usual doldrum weather, until 31st October, when she was in 31° N., 33° W. Here strong northerly gales were encountered, and she was under small sail for a whole week with high seas and heavy weather.

On 6th November her abstract log read as follows:—"A.M., gale decreasing, wind N.N.W., set reefed topsails. 10 a.m., wind and sea increasing. Ship making some tremendous plunges and smothering herself with water. Laid her to the wind under lower topsails. Noon, lat. by account, 35° 32′ N., long. by account, 26° 50′ W.; course, N. 81° E.; distance 107 miles. Whole gale and very heavy sea. Shipping some very heavy lumps. 8 p.m., gale increasing, and was going to take in fore and mizen topsails when she took a plunge and broke the jibboom. Kept her away to clear the wreck, let her run under lower topsails. Wind N.W. and westering. Midnight, whole gale with rain. Tremendous northerly sea wind N.W. by W.

At 1 p.m. on 16th November, *Norman Court* made the Lizard in a fresh north-westerly gale; and she took her pilot off Dungeness just 24 hours later. That night she anchored in the Downs, 113 days from Foochow. *Sir Lancelot* arrived in the Downs on the following day, 18th November, so that *Norman Court* beat her home by one day from the Formosa Channel. Neither vessel was favoured by the weather, either in the China Sea, off the Cape or in the North Atlantic, or their times would have been much better.

In 1875 *Norman Court* came home from Shanghai in 120 days, leaving 2nd September and on her next voyage did the same trip in 106 days with the favourable monsoon. Then in 1877 the depression in freights led her owners to take the usual steps, and, whilst she was on the China Coast, orders came out for the yards to be stripped off her mizen mast and her crew to be reduced. Hardly was this done, however, before she was chartered to load tea at Hong Kong for Port Elizabeth, South Africa—the tea being sent down to Hong Kong by steam. And her charter contained a clause which stated that she was to get 5/- a ton extra if she arrived in Port Elizabeth before a German clipper barque similarly loaded.

The German got away 10 days ahead, but in spite of a slow passage to Anjer in the month of August, *Norman Court* arrived at Port Elizabeth on 28th September, 61 days out and 20 days before the German.

She then loaded wool at the Cape for London and on 19th January, 1879, made the Scillies, 43 days out from Table Bay. It was that terrible January when the Thames was frozen over, and old seamen will also remember it for the hard time they had in the Channel battling against the chilly blast of the N.E. gale, which seemed as if it would never end.

Norman Court came in for the full brunt of it, and spent three days off the Wolf Rock beating under lower topsails; unlike most of the homeward bounders, she did not run into Falmouth as soon as she was round the Lizard, but battered her way up Channel, tack and tack. In the cabin Captain Shewan and his passengers strove to keep warm with Cape brandy punch and sea pies, which had to be brought aft in the saucepan and eaten on the cabin settee, so heavily was the *Norman Court* pitching into it.

The crowd forward, however, with no such luxuries, had 12 days of icy spray, wet clothing and "maintopsail haul," and, at last, when a tug bore down on the hard-used ship off Beachy Head, they listened to the bargaining with the greatest of anxiety.

The wind had moderated by this time, though still a "dead muzzler," and *Norman Court* had all sail set. The tugboat man wanted a big sum to tow the clipper to London. Captain Shewan offered him £50, at which the tug sheered off, and Captain Shewan immediately went about and stood out to sea again.

Again the tug came alongside. "£60," he roared. But Captain Shewan would not give in and once more the tug dropped astern. Once more the *Norman Court* was filled away, her crew literally groaning as they hauled aft the main sheet. The sail was hardly set, however, before the steamboat came panting up again, and the welcome cry rang out across the short Channel sea, "Haul that mainsail up and give us your rope."

With a stentorian cheer the *Norman Court's* crew flew to the buntlines and clew garnets and the ship was stripped of her canvas in record time.

The *Norman Court* was the only ship to beat up Channel in that month of freezing easterly gales, with the exception of the beautiful wool clipper *Mermerus*, one of the finest and fastest iron ships ever built. Both ships, however, might have saved the January wool sales, if they had put into Falmouth instead of keeping the sea, for so bad was the weather and so great was the fleet in Carrick Roads that special tugs were chartered at cheap rates to tow the delayed shipping up to the London River.

After this strenuous finish to many years of tea racing, Captain Shewan was so worn in health that he decided to take a rest and Captain Dandy Dunn took the *Norman Court* out in 1880 and brought her home with what was to be her last tea cargo. On her next voyage she went to the Coromandel Coast, and then Barings sold her to a firm in Glasgow for the Java trade.

On her first homeward passage under her new owners, she was running up the Irish Channel before a stiff sou'wester when she got hard and fast ashore at the back of Holyhead and went to pieces. Her beautiful figurehead is still preserved in a garden near Holyhead.

I must now turn to *Hallowe'en.* Her record in the China trade was a truly wonderful one for an iron ship, though it must be remembered that she always sailed late and had the advantage of the favourable monsoon; the sailing and arrival dates of her five best passages were:—

1873—Shanghai to London,	Nov 19 to Feb. 16	..	89	days
1874— ,, ,,	Oct. 21 to Jan. 19	..	90	,,
1875— ,, ,,	Nov. 23 to Feb. 23	..	92	,,
1876— ,, ,,	Nov. 13 to Feb. 23	..	102	,,
1881— ,, ,,	Nov. 27 to Mar. 10	..	103	,,

I give the log of her 1873 passage in the appendix. Those of 1874 and 1875 show the same uniformly good times between the various points. In 1875 she was delayed two days by having to stop at St. Helena in order to land Captain Watt, who was so ill that he died almost before his ship was out of sight of the island.

His chief officer Fowler, brought the *Hallowe'en* home and had command of her for the next few years. Her outward passages were usually made to Sydney, and though she never equalled her maiden passage she was considered the only vessel which could seriously rival *Thermopylae* and *Cutty Sark* in speed.

On 19th August, 1886, she left Foochow for London with what was to be her last cargo of tea, for at 7.30 p.m. on 17th January, 1887, she ran ashore near Salcombe and became a total wreck.

Lothair, the last of the tea clippers (I do not count the iron *Serapis,* which was not launched until 1875 and saw very little of the tea trade) was converted to a barque in the early eighties, and for another ten years she still flew the Red Ensign. Then she was sold to the Genoese owners who again sold her to the Peruvians about 1906 or 1907, and she did not disappear from the register until 1911; indeed it is quite possible that she is still knocking about the South Pacific with Callao as her home port.

We now come to the great *Thermopylae.* She remained in the service of her original owners, the famous Aberdeen White Star Line, until 1890, and the times of her outward and homeward passages whilst under their house flag (which are given in the appendix) show the wonderful consistency of her work, in spite of the fact that her sail and spar plan was twice reduced. The average of her best ten passages out to Melbourne, from pilot to pilot, give the astonishing time of 67 days, and Captain Jenkins, her last commander under Thompson's flag declared that on the 31st December, 1888, only two years before she was sold, she made 358 miles in the 24 hours, whilst running her easting down in 44° S., 68° E., bound out to Sydney.

Her only bad passage out to the Colonies was the one to Sydney in the winter of 1882-3. But the following abstract from her log will easily explain the reason of this:—

Date	Lat.	Long.	Course	Dist.	Remarks
1883	° ′	° ′	°		
Jan. 22	—	—	—	—	Light easterly winds 2 p.m. Portland bore N.W. by W. Landed Mr. Cobley, pilot.
23	49 21 N.	5 2 W	W by S.	115	Mod. S.S.E. winds. Heavy head sea.
24	48 53	7 50	S. 77 W.	129	Fresh southerly breeze and thick heavy confused sea.
25	48 2	8 24	S. 12 W.	50	Strong westerly breeze.
26	46 34	9 16	S. 22 W.	95	Strong westerly gale. Heavy squalls. High sea. Ship under lower topsails.
27	45 42	9 16	South	52	Strong gale and heavy squalls. Shifting from N.W. to west. Head reaching on both tacks.
28	46 10	10 08	N. 53 W.	48	Strong gale from west. Heavy squalls.
	° ′	° ′	°		
Jan. 29	45 58	10 02	S. 14 E.	15	Ship laying to with heavy W.S.W. gale and rain. 10 a.m. shifted to N.W. Heavy sea. Ship labouring very much.
30	44 51	10 34	S. 26 W.	75	Strong N.W. gales and heavy confused sea. Ship labouring heavy and shipping much water on deck.
31	45 4	10 38	N. 17 W.	8	Violent gale from S.S.W. to N.W., very heavy sea. Main lower topsail and new mizen staysail blew to pieces.
Feb. 1	44 45	10 02	—	—	Violent gale, ship laying to under fore-topmast staysail and lee clew of fore-topsail. 8 p.m., fore topsail blew away.
2	45 9	9 50	S. 34 W.	14	Violent gales veering from W.S.W. to N.W.
3	44 28	10 04	S. 15 W.	44	Strong W.N.W. gale and heavy confused sea. No lower topsail set. Sailing by the wind.
4	44 56	11 26	N. 65 W.	65	Fresh westerly breeze and clear. Bending lower topsails; wind freshening. Sailing by the wind.

Date	Lat.	Long.	Course	Dist.	Remarks
Feb. 5	45 02	12 26	N. 84 W.	41	Steering by the wind; strong S.W. breeze. Ship under lower topsails and staysails.
6	45 10	12 51	N. 66 W.	20	Strong S.S.W. gale and rain. Midnight calm. Noon, strong S.W. gale, ship laying to.
7	44 24	13 51	S. 45 W.	67	First part strong S.W.; second part strong N.W. gale.
8	42 46	14 42	S. 23 W.	108	Strong N.W. breeze and clear up to midnight, then strong W.S.W. gale to noon.
9	42 53	15 19	N. 73 W.	30	Strong W.S.W. gale for 16 hours, then shift to N.W. Ship head reaching.
10	42 46	14 42	S. 22 E.	28	Strong W.S.W. gale; ship head reaching and laying to.
11	41 38	14 45	N. 10 W.	45	Heavy W.S.W. gale and rain, latter part strong gale.
12	39 26	14 01	S. 8 E.	124	Strong W.S.W. breeze and clear. Reefed upper topsails. Wearing ship.
13	39 6	13 38	S. 36 E.	24	Strong gale from W.S.W.
14	—	—	—	—	Strong gale from W.S.W.
15	38 35	14 06	S. 35 W.	38	First part fresh W.S.W. breeze and rain, latter part light airs and calms.
16	37 08	15 13	S. 32 W.	102	Light southerly winds and calms. Heavy N.W. swell.
17	36 53	16 06	S. 71 W.	45	Calms and variables. Tacking.
18	36 18	15 59	S. 9 E.	35	Light variable airs.

At last on the 19th the sorely battered *Thermopylae* took the N.E. trades. She was actually 45 days from Portland to the line. On her previous voyage she had crossed the equator 16 days out from the Lizard Light and 17 from where she landed her pilot off Dartmouth.

I can only find one instance of her ever being passed at sea except by *Cutty Sark*, and that was when homeward bound from Sydney in 1883, when I find the following entry in her log:—"31st December. Lat. 6° 0′ S., long. 29° 48′ W. Course, N. 28° W. Distance 101 miles. Wind N.E., steering by the wind (wind light). Spoke a German barque which went right out ahead of us in 24 hours from S.W. to N.N.E. There is no mistake but she gave us the go by in style—the first I have seen do so."

Perhaps *Thermopylae's* copper was ragged on this occasion, as the usual entry was more like this:—"A.M., ship in sight on weather bow, going same way. P.M., same ship hull down on lee quarter."

Thermopylae was the pride of Thompson's fleet, and it must have been with great reluctance that they sold her in 1890 to Mr.

Reford of Montreal, President of the Rice Milling Company. The latter took off her the Australian route and put her into the rice trade between Rangoon and Vancouver, British Columbia.

Even in these last years she made many fine runs, notably one of 29 days between Shanghai and Victoria, B.C.

Her last passage under the British flag was made between Port Blakely and Leith, over which she took 141 days. She was then (1895) sold to the Portuguese Government, who turned her into a training ship and renamed her the *Pedro Nunes*.

Finally on the 13th October, 1907, she was towed out of the Tagus by two Portuguese men-of-war and torpedoed. Some people stated that she was simply used as a target in a naval display, others that the Portuguese Government, finding that she was too old and too small for the service on which she was engaged, decided to give her a "naval funeral" in honour of her splendid achievements in the past, and therefore ceremoniously towed her out to sea and sunk her with colours flying and bands playing.

Let us hope that it was this latter most worthy sentiment that inspired the Portuguese.

I have left the *Cutty Sark* to the last for the simple reason that she is the only ship out of the whole tea fleet which is still afloat—and not only afloat but still earning her living on the great waters.

In the appendix I give a complete record of her wonderful work in the Australian trade whilst under the British flag. A few notes on the best of her passages will, however, be of interest.

In the winter of 1877-8 she made her best passage out to Sydney, after very nearly ending her days on the Goodwin Sands during one of the severest gales of the century.

She had left London with a general cargo for Sydney on 3rd November, 1877. But the weather was so bad in the Channel that after fighting vainly for five days to make headway against the gale, she put back to the Downs, and Sunday, the 11th of November, found her riding out a furious S.W. gale with 105 fathoms on her starboard and 60 fathoms on her port anchor, whilst another 200 fathoms were ranged on deck, in case of emergency.

The storm grew steadily worse until at 10.35 p.m. both cables parted the wind now blowing a hurricane.

Before sufficient sail could be set to put the *Cutty Sark* under control of her helm, she ran foul of two ships anchored near her, first crashing into a brig on her port and then a ship on her starboard hand. Luckily, however, she drifted clear before any very serious damage was done.

A very heavy sea was running and the night was pitch dark. Captain Tiptaft with a crew of 28 all told managed to set a fore topmast staysail, reefed foresail and main lower topsail and steered a course through the Gulf Stream, then as soon as he was to the nor'ard of the Goodwins and about 8 miles north of the North Foreland Light, he hove her to on the starboard tack with her head to the southward.

Cutty Sark was still, however, in very grave danger both by reason of the furious gale and the damage she had already sustained. She had no anchor available or which could be got over the bows in such a sea as was running and only 45 fathoms of chain remaining.

The foresail had blown away along with other sails torn out of their gaskets, the lower topsail had split and only the fore topmast staysail remained intact. At the same time, owing to the two collisions, all her port fore braces were gone, so that the yards could not be braced up, and, lastly, her bulwarks had been levelled to the neck.

Not knowing what other damage might not have been sustained and taking into consideration the helplessness of his position, Captain Tiptaft now decided to send up rockets and burn flares for assistance. At about 4.30 p.m. these attracted the attention of the tug *M'Gregor*, which was some distance to the S.W. of the Kentish Knock. The tug at once set off at full speed and within an hour was alongside the distressed vessel, which had by this time drifted close to the South Kentish Knock buoy.

With great difficulty the *M'Gregor* managed to pass her rope aboard, but when it came to towing, it was found that with her utmost pressure of steam, 27 lbs., she could only progress at the rate of a mile an hour, and by 10 a.m. she had only towed the *Cutty Sark* 6 miles. However, another tug, the *Benachie*, now got hold, and between the two of them the hardly-used ship was brought safely through the Princes Channel, and at 7 p.m. was safely moored at Greenhithe. For this night's work the tugs were awarded £3000, the value of the *Cutty Sark*, her cargo and freight being agreed at £85,000.

After being repaired and re-fitted, *Cutty Sark* made a fresh start, and, taking her departure from the Lizard on 6th December, made the best passage of the year to Sydney, arriving 16th February, 72 days out.

From Sydney she went across to China, but Captain Tiptaft found the homeward rates so bad that he was glad to load tea at Hankow back to Sydney. On her way down the Yangtse the *Cutty Sark* had the misfortune to lose two anchors and chains and break her port hawse pipe, and on her arrival at Shanghai her captain died. He was succeeded by a Captain Wallace, who took the *Cutty Sark* across to Sydney and back and then loaded at Manila for New York.

Captain Wallace was a sail carrier, and brought the *Cutty Sark* across to London from New York in 10 days, 18 hours of which were spent hove-to bending new sails. In May, 1880, after having had her sail plan cut down, *Cutty Sark* loaded coal at Cardiff for the American Squadron in the East. A very profitable charter was the reason for old Willis running his clipper into a collier—but he bitterly repented this insult to his ship.

The voyage was a tragedy and a financial failure, from start to finish. Down in the roaring forties, the mate, a notorious Yankee bucko, killed a man with a handspike. On arrival at Anjer Captain Wallace allowed the mate to escape to an American ship. This so enraged the crew that they refused to get under weigh. However, the apprentices and idlers, who were all devoted to Captain Wallace, manned the capstan and got the ship to sea. Then for two whole days the *Cutty Sark* lay in a clock calm; Captain Wallace had not slept for a week owing to the worry over the mate, and one morning watch he slipped over the stern and was taken by sharks before a boat could be got over. The second mate thereupon put back to Anjer, where a Dutch pilot was obtained, and the ship taken through Banka Strait to Singapore. Here an inquiry was held and the crew discharged. The mate of the *Hallowe'en* was sent down from Hong Kong to take command. This man, named Bruce, converted the poor little *Cutty Sark* into a hell ship. After a protracted round of Calcutta, Melbourne, Sydney, and Shanghai, during which everything went wrong that could go wrong, and sailormen were chased out of the ship to make way for shilling-a-month men, the ship eventually turned up in New York from Cuba, spring of 1882, having had to cadge provisions from passing ships three times on the passage.

Reliable old Captain Moore was immediately sent out from England to supersede the incompetent and rascally Bruce.

Moore had the *Cutty* for three voyages, but he was growing old and no longer carried sail, so it was not until he was succeeded by Captain Woodget in the spring of 1885 that *Cutty Sark* was once more allowed to show her paces.

On her first outward passage under Captain Woodget she had a most interesting race to Sydney against the two iron wool clippers *Samuel Plimsoll* and *Sir Walter Raleigh*, and also one of Smith's fast cities, the *City of York*.

All three vessels were noted for their speed, and the result was of the most level description as the following times show:—

Ship	Off the Start	Crossed the Line	Passed Cape Meridian	Arrived Cape Otway	Arrived Sydney	Days Out
City of York ..	April 2	April 23	May 26	June 18	June 21	80
Cutty Sark ..	„ 3	„ 23	„ 19	„ 16	„ 20	78
Samuel Plimsoll ..	„ 4	„ 28	„ 21	„ 18	„ 21	78
Sir Walter Raleigh	„ 4	„ 28	„ 22	„ 20	„ 22	79

Cutty Sark's first three homeward passages from Sydney under Captain Woodget came very near being a record.

On the first she was off the Lizard 67 days out.

On the second she left Sydney 26th March, 1887.

Passed Cape Horn 21st April	.. ..	26 days out.
„ Equator 13th May	.. .. ..	48 „
„ Azores 25th May	.. .. ..	60 „
Took her pilot off Dungeness 6th June	..	72 „

On her third she was off Brighton at 1 p.m. on 8th March, 71 days out.

But right up to 1895 she continued to do good work. Unlike the American clippers, which lost a great deal of their speed in old age through becoming water-soaked, *Cutty Sark*, when over twenty years old seemed to be as fast as ever. In 1895 she was sold to the Lisbon firm of Ferreira & Co., her name being changed to *Ferreira*, yet only a few years back I heard that she could still do her 16 knots without much fuss. For the past few years she has been making one leisurely voyage a year, half the time being spent in port, her round being usually Lisbon to Rio de Janeiro, then New Orleans, and from there back to the Tagus. Though, of course, the Portuguese make no attempt to drive her, they have kept the yards on the mizen and she makes very good and regular passages.

An officer of the Mercantile Marine, who saw her in New Orleans last May, has sent me the following interesting account of the old ship:—

"Strolling leisurely one day along the water front at New Orleans, I noticed standing prominently out behind an old shed the tall tapering spars of a sailing ship. This class of cargo carrier being more the exception than the rule at the wharves of the Crescent City, and taking as I do a keen interest in the doings of old clippers, my curiosity tempted me to investigate, so retracing my steps I made the best of my way through a timber yard and eventually emerged upon the old and dilapidated wharf at which she lay. The day of clipper ships was past and gone in long ere I commenced my apprenticeship in a modern Clyde four-poster, but I needed no telling that this was one of the old timers.

"The sun, high in the heavens, shone down with a dazzling glare on her weather-beaten hull, painfully emphasising every detail of its shabby exterior and general air of neglect, but though shorn of much of her former glory the unmistakable stamp of an aristocrat of the sea was ineradicable. It shone forth despite her tattered gear and pitted bulwarks. Like the old racer one sometimes sees relegated to the 'shafts', the breed was unmistakable.

"Floating lazily aloft with the shield and crown of Braganza's noble house graven upon it, was the ensign of Portugal. Wondering vaguely what old clipper she might be, I sauntered along the wharf admiring her graceful lines. She was shiprigged with single top-gallant sails and composite built. Her copper sheathing was visible apparently intact. Looking at her from forward, her entrance was like the thin edge of a wedge and it filled out gradually to her waist. A little fuller perhaps in the run, she had a handsome stern, whilst blazoned on her deep counter in 6-inch yellow letters was the legend 'Ferreira * Lisboa.'

"For a figure-head she had a comely maiden with swelling bosom and hand outstretched pointing ahead—plentifully bedaubed with multi-coloured paint. Though in hopes of finding some trace of her old name on the bows I searched in vain—everything was obliterated and only the glaring *Ferreira* remained.

"Making up my mind to go aboard, I glanced round to see if there was anyone in authority, whose permission ought to be asked. Everybody in the vicinity seemed to be enjoying their siesta. Several huge piles of staves, her cargo, lay around, upon which sundry 'niggahs' lazily basked, whiling the sunny hours as only a Southern nigger can. Walking over the gangway, I made my way slowly aft and mounted the poop.

"To give the dagoes credit, they certainly did devote a little attention to this part of the ship though occasional startling splashes of colour (so dear to the Portuguese) struck a jarring note. The upper poop consisted of a raised deckhouse, some 3½ feet high. It was neatly railed and hammock-netted round. Along the port and starboard sides ran a row of garden seats. I call them garden seats as they were of a pattern more generally found in parks and gardens than on board ship. Two individuals occupied this poop, one worked away, stitching on the gore of a topsail, the other slumbered peacefully on one of the garden seats. The running gear all came down to the outer or lower poop, from which the mizen rigging was set up. Walking round this outer poop I came to the after end of the upper one, abaft which was the steering gear.

"I examined the wheel and gear with interest, and also the brass bell, but though both were of an old pattern, I failed to find any trace of the ship's original name on either. Advancing on the individual who was goring the topsail and who, by the way, did not seem in the least disturbed at my presence, I addressed him—

" 'You speak English?'

" 'He looked up and shook his head.

" 'Are you an officer?' I hazarded.

" 'No, sabe.'

" 'Where is the captain?' I asked at last as a sort of forlorn hope. The reply somewhat astonished me.

" 'Me captain,' he said, and went on with his work. I then made various gestures to signify I would like to see down below. He nodded acquiescence, so leaving him to his stitching and the 'una pelota' (for such I took him to be) to his slumbers, I descended the after companion.

"An alleyway led into the saloon on either side of which were doors with cut glass handles. The saloon was a fairly spacious apartment running athwartship. It was panelled neatly in teak and bird's eye maple and was adorned with much fancy carving. Beautiful as it had evidently once been, it was pretty bare now, the marble-topped sideboard and fireplace and the old brass lamp which swung in the skylight being probably the only original furniture left.

"Another alleyway led from the saloon forrard, and as I passed along it I glanced in through an open door into the captain's room. Like the saloon it was stripped of most of its old fittings, only a marble-topped washstand and a heavy, teak four-post bed (the latter not often seen in ships nowadays) remaining. Various rooms occupied either side of the alleyway and at the end another companionway gave egress to the lower poop. Not caring to intrude I investigated no more of the rooms beyond noticing over the doors that old familiar legend 'certified to accommodate one seaman.'

"Entering a door under the companion stairway I presently found myself in the after 'tween deck. Overhead the rust clung in huge scales to the diagonal tieplates and beams. The frames by the feel of them were still in a fair state of preservation, though they had not known a hammer or slice or paint for many a day.

"Along the port and starboard sides ran a row of ports (now all plugged up) showing that at one time she had carried human freight here—emigrants, no doubt.

"Coming to the main hatchway, I peered closely at its pitted surface endeavouring to decipher some letters and figures cut on the after coaming, but only managed to make out 63556 and $921\frac{}{100}$ tons.

"Continuing forward through the fore 'tween decks which contained the usual miscellaneous collection of old junk, blocks, and rusty wire, I came to the fore hatch. And as I looked down below at her wedgelike entrance, I thought that assuredly it needed clipper freights to make the ship pay. One could hardly find room to stand up on either side of the keelson, so fine was she. The iron collision bulkhead came down triangle-shaped, the apex at the keelson, and I mentally compared it with those of some modern windjammers and tramps which form nearly a square.

"Retracing my steps aft and climbing through the after hatchway, I reached the deck again and was not sorry to feel the bright sunshine, for the old 'tween deck had a chilly, eerie atmosphere about it.

"Gazing round, I now found many things to interest me. Her decks were badly rutted and cracked and sorely needed oil. Her rigging, fitted with wire lanyards (a doubtful boon) would have been better for a little tar and service. As the Yanks would say, they were 'Hell on chains,' chain strops being in abundance. And where a backstay had parted or a fore and after gone in the nip, the deficiency was supplied in this manner. A very handy device caught my eye abaft the main rigging, viz., a single winch barrel with double purchase and handle clamped to the topgallant pin rail. Apparently it could be used with equal facility for taking in a bit on the main sheet or bousing down the crossjack tack in a stiff breeze. It did not look, though, as if it had had much use of late.

"The teakwood stanchions at the break of the poop, once a mass of shining brass and glistening varnish were now—ye gods! —painted with aluminium paint. It would have made an old deep-water mate grind his teeth to see such a desecration of the time-honoured methods of preserving 'bright work.' Near by a row of teakwood buckets stood in racks. These were brown painted and adorned with silver bands—too much trouble to scrub them, I suppose.

"As I walked past I could not help glancing in to what had once been the half-deck. The door was open, so seeing no one at home I stepped in. A roomy enough place, it apparently once provided accommodation for quite a number of apprentices. It was now the abode of the petty officers; its old deal table, well-worn floor and battered bunks quite reminded me of old times. In the fore part of the after house a donkey room was situated containing an engine and winch of ancient pattern. Overhead were the boat skids upon which two launches and two boats rested in chocks, whilst on the deck above the old harness casks were still in possession.

"The main fife-rail, inside of which the original old bilge pumps stood, was in pretty bad shape, though it must at once have looked very fine with all its brass and carving. 'Way up above the rail I noticed the lower block of the topsail halliards, a chain pennant reached from it to the deck, a rope-saving device no doubt. The forward house, a neatly panelled structure, was identical with the half deck but somewhat larger. She was well provided with boats, two more being on the top of this house—a wise precaution, as some day, like the 'one horse shay' she will go to pieces all at once.

Making my way up the ladder, I reached the foc's'le-head, a pretty bare spot, enclosed by sundry rust-eaten stanchions with a ridge rope rove through them. Two pairs of hardwood bollards were placed on each side, on one of which a solitary brass cap glistened forlornly. The old whisker-booms were still in use, one out, the other in, all askew. The jibboom was rigged in, and as I looked at the old spar the lines of a deep-water song came to my mind.

" 'There was no talk of shortening sail by him who trod the poop,
And her boom with the weight of a mighty jib bent like a wooden hoop.'

"Looking over the side I again admired her clean entrance and knife-like bows. The old wooden-stocked anchors hung at the cat-heads, and the ring stoppers were fitted with a patent 'tumbler' releasing gear eliminating the use of the time-honoured maul.

"Coming down from the foc's'le-head, I had almost made up my mind to go when something caught my eye, which I had overlooked. Standing in pathetic solitude, suspended from a solitary cast-iron dolphin, was the old forward bell. Surely this would give me a clue to her name, I thought. I went up and examined it closely.

Its surface appeared at first sight perfectly smooth, thickly coated with silver paint as it was. Presently, however, I thought I could discern a very faint trace of lettering. At which I extracted my knife and scraping away gently, gradually revealed the date 1869. I now hesitated, not caring to further mutilate the Portuguese artistic work but reflecting that I might as well be hung for a sheep as a lamb, I took up my knife again. A few more strokes of the sharp blade and there, standing out boldly was a name, once a byeword amongst seafarers, which raised a thrill such as that of the *Mauretania* or *Lusitania* could never raise—the *Cutty Sark*! I tapped the old bell gently with my knife and heard again that mellow sound which through the trades, the tropics and the roaring forties had for nigh half a century marked alike the dark and the sunny hours.

"Well, time was flying, and I had a long walk before me, so I made my way ashore. Standing on the wharf I surveyed her once more with a keener interest. The setting sun had almost reached the horizon. Its mellow, golden light, shining on her spars and rigging, seemed to transform her and clothe her in some of her ancient glory. Hidden were the marks of decay, and she once more looked the ship of speed and beauty."

It is evident from this account that the famous old ship has nearly run her time, though my friend says her copper was in good condition and that he could see no water in her hold. It is a curious coincidence that the two great rivals, *Thermopylae* and *Cutty Sark*, should have spent their old age under the same flag and with the same home port.

It has always been a wonder to me that some rich deep-water yachtsman did not think of picking up one of these beautiful little tea clippers when her racing days were over, and converting her into a yacht. Such a vessel as *Thermopylae*, *Cutty Sark*, *Sir Lancelot*, *Lothair*, *Leander* or *Titania* could have been had for a song in the early nineties—and what a yacht she would have made! All that would have been needed would have been a couple of launches on the skids and the conversion of the hold into a spacious suite of cabins. Such a yacht would have been a worthy flagship for the R.Y.S., bringing back memories of Lord Yarborough's famous *Falcon*.

If we had not been a nation of shopkeepers there is little doubt but that the *Thermopylae* or *Cutty Sark* would still be flying the British flag, preserved by public subscription for the important part they played in the greatness of our Mercantile Marine.*

This book has been a small attempt to preserve the records of these beautiful ships before their history and achievements are forever forgotten, and I bring it to a finish with the sad thought that along with the China clipper there has departed not only that high art called seamanship, but also much of the romance, charm and virility of sea life.

In 1922 a retired windjammer skipper, Captain Dowman, of Trevissome, near Plymouth, bought the *Cutty Sark* from her Portuguese owners paying £3750 for the old ship. Captain Dowman re-rigged the "Cutty" according to her original sail and spar plan, and dreamed of her running her easting down and making the heads of Port Jackson in 70 days. But Captain Dowman died, and in 1938 his widow, who took as great an interest in the old clipper as had her husband, presented the "Cutty" to the Thames Nautical Training College. She can be seen moored abreast of the *Worcester* off Greenhithe.

APPENDIX

APPENDIX A.—*British Tea Clippers.*

Date Built	Name of Ship	Best Known Commander	Tonnage	Length	Breadth	Depth	Builders	Owners
1850	*Stornoway* ..	Robertson	506	157′ 8″	28′ 8″	17′ 8″	Hall, Aberdeen	Jardine, Matheson
1851	*Chrysolite*	Enright	471	149 3	26 1	17 0	Hall, Aberdeen	Taylor & Potter
1852	*Challenger* ..	Killick	699	174 0	32 0	20 0	Green, Blackwall	Lindsay
1853	*Cairngorm*	Robertson	938	185 0	36 6	20 0	Hall, Aberdeen	Jardine, Matheson
,,	*Northfleet*	Freeman	896	180 0	32 3	20 9	Northfleet	Duncan Dunbar
,,	*Lord of the Isles* ..	Maxton	770	190 9	27 8	18 5	Scott, Greenock	Martin
,,	*Crest of the Wave*	Steele	924	184 0	32 3	20 1	Pile, Sunderland	Brice
1854	*Spirit of the Age* ..	Billing	878	173 0	32 0	18 5	Pile, Sunderland	T. Gibb & Co.
1855	*Fiery Cross* (1) ..	J. Dallas	788	154 5	31 0	19 1	Chaloner, Liverpool	J. Campbell
,,	*Kate Carnie* ..	Rodger	576	148 4	26 0	19 0	Steele, Greenock	Rodger
1856	*Robin Hood* ..	Cobb	852	204 0	35 1	21 0	Hall, Aberdeen	Beazley & Co.
,,	*Lammermuir* ..	Shewan	952	178 0	34 0	22 0	Pile, Sunderland	J. Willis
1857	*Friar Tuck* ..	Fordyce	662	193 2	31 0	17 0	Hall, Aberdeen	Beazley & Co.
1858	*Ellen Rodger* ..	Keay	585	155 8	29 4	19 5	Steele, Greenock	Rodger
,,	*Ziba*	Tomlinson	497	169 0	28 5	17 0	Hall, Aberdeen	J. Wade
1859	*Falcon*	Maxton	794	191 4	32 2	20 0	Steele, Greenock	Shaw & Co.
1860	*Fiery Cross* (2) ..	Robinson	695	185 0	31 7	19 2	Chaloner, Liverpool	J. Campbell
,,	*Flying Spur* ..	Ryrie	735	184 0	31 4	19 4	Hall, Aberdeen	Jardine & Co.
,,	*Lord Macaulay* ..	Care	846	168 0	35 0	21 5	Hall, Sunderland	Brodie
1861	*Highflyer*	Enright	1012	193 7	35 5	20 0	R. & H. Green	R. & H. Green
,,	*Min*	Smith	629	174 5	29 8	19 3	Steele, Greenock	Rodger
,,	*Kelso*	Vowell	556	150 0	31 3	18 5	Pile, Sunderland	J. Kelso
,,	*Whinfell*	Jones	834	190 2	32 6	22 2	Lamport	Lamport
,,	*Silver Eagle* ..	Case	903	185 2	34 5	20 8	Portland Ship Co.	Joseph Soames
1862	*White Adder* ..	Moore	915	191 4	34 0	20 7	Bilbe, London	J. Willis
,,	*Guinevere*	M'Lean	646	175 0	29 9	19 3	Steele, Greenock	J. MacCunn
,,	*Coulnakyle* ..	Morrison	579	168 0	30 5	18 8	Hall, Aberdeen	Jamieson
,,	*John Lidgett* ..	Polson	770	178 7	30 1	20 4	Stephen, Glasgow	Lidgett
,,	*Star of China* ..	Hodge	794	183 0	32 0	20 2	Hall, Aberdeen	Adamson
,,	*Vigil*	Thomson	550	163 9	27 2	18 0	Vernon	Potter
,,	*Burdwan*	Douglas	803	185 7	32 1	21 6	Brocklebank	T. & J. Brocklebank
1863	*Taeping*	M'Kinnon	767	183 7	31 1	19 9	Steele, Greenock	Rodger
,,	*Serica*	Innes	708	185 9	31 1	19 6	Steele, Greenock	Findlay
,,	*Belted Will* ..	Braithwaite	812	186 4	32 4	20 8	Feel, Workington	Bushby
,,	*Eliza Shaw* ..	Steele	696	184 5	30 6	18 3	Stephen, Glasgow	Shaw, Maxton
,,	*Pakwan*	Shiel	795	186 0	32 5	19 0	Peverill, Sunderland	Patton
,,	*Black Prince* ..	Inglis	750	183 0	35 0	19 6	Hall, Aberdeen	Baring
,,	*Fychow*	Mathers	710	180 0	31 5	19 2	Hall, Aberdeen	Dunbar
,,	*Elizabeth Nicholson*	Grierson	904	192 5	32 5	22 2	Nicholson	Nicholson
,,	*Dunkeld*	Toms	699	170 0	30 3	19 3	Duthie, Aberdeen	Foley
,,	*Wild Deer* ..	Smith	1126	211 0	33 2	20 7	Connell, Glasgow	Albion Co.
,,	*Red Deer*	Mills	775	160 0	32 5	19 7	Barr	Adamson

APPENDIX A.—*British Tea Clippers—Continued*

Date Built	Name of Ship	Best Known Commander	Tonnage	Lsngth	Breadth	Depth	Builders	Owners
1863	*Roslyn Castle*	Nicol	644	183′ 9″	29′ 1″	18′ 6′	Connell, Glasgow	Skinner
,,	*Yangtze*	Kemball	688	179 5	31 0	18 3	Hall, Aberdeen	Lewin
1864	*Dilkhoosh*	Gedge	816	167 2	32 4	21 2	Major	Fleming
,,	*Dilpussund* ..	Jones	624	180 0	29 0	18 0	Langley	Fleming
,,	*Douglas Castle* ..	M'Kitchie	678	176 6	30 6	18 7	Connell, Glasgow	Skinner
,,	*Gossamer*	Thomson	735	181 4	30 6	18 4	Stephen, Glasgow	Potter
,,	*St. Andrews' Castle*	M'Bain	639	168 8	30 3	18 2	Connell, Glasgow	Skinner
,,	*Golden Spur* ..	—	657	177 4	31 4	19 9	Ogier, Guernsey	G. T. Carrington
1865	*Ariel*	Keay	852	197 4	33 9	21 0	Steele	Shaw, Maxton
,,	*Sir Lancelot* ..	Robinson	886	197 6	33 7	21 0	Steele	J. MacCunn
,,	*Taitsing*	Nutsford	815	192 0	31 5	20 1	Connell	Findlay
,,	*Chinaman* ..	Downie	668	171 0	31 1	19 1	Steele	Park Bros.
,,	*Ada*	—	687	182 0	30 0	18 0	Hall	J. Wade
,,	*Maitland*	Coulson	799	183 0	35 0	19 6	Pile	Kelso
1865	*J. R. Worcester* ..	Wawn	844	191 5	32 4	19 9	M. Inv. Co.	Patton
,,	*Fusiyama* ..	Thomson	556	165 6	28 1	17 0	Stephen	Killick
,,	*Lennox Castle* ..	Brunton	693	178 6	30 1	18 9	Moore	Skinner
1866	*Titania*	Dowdy	879	200 0	36 0	21 0	Steele	Shaw, Maxton
,,	*Sir W. Wallace* ..	Taylor	967	195 6	34 5	21 1	Duthie	Tulloch
,,	*Huntly Castle* ..	Stewart	623	169 0	29 0	18 2	Connell	Skinner
,,	*Wemyss Castle* ..	Nicol	700	183 0	31 0	17 9	Connell	Skinner
1867	*Spindrift*	Innes	899	219 4	35 6	20 2	Connell	Findlay
,,	*Lahloo*	Smith	799	191 6	32 9	19 9	Steele	Rodger
,,	*Leander*	Petherick	883	210 0	35 2	20 8	Lawrie, Glasgow	Joseph Somes
,,	*Undine*	Scott	796	182 6	35 1	19 5	Pile	J. Kelso
,,	*Forward Ho* ..	Hossack	943	193 7	33 6	20 6	Stephen	Catto
,,	*Kinfauns Castle* ..	Holmes	799	187 6	32 4	19 4	Connell	Skinner
1868	*Thermopylae* ..	Kemball	948	212 0	36 0	20 9	Hood, Aberdeen	George Thompson
,,	*Windhover* ..	Nutsford	847	201 1	34 p	19 8	Connell	Findlay
,,	*Kaisow*	Anderson	795	193 2	32 0	20 3	Steele	Killick
,,	*Omba*	Thomson	836	186 6	31 8	19 5	Stephen	Killick
,,	*Carrick Castle* ..	Peters	879	197 5	34 0	19 6	Elder, Glasgow	Skinner
1869	*Cutty Sark* ..	Moodie	921	212 5	36 0	21 0	Scott & Linton	J. Willis
,,	*Wylo* ..	Brown	799	189 9	32 1	20 2	Steele	Hillick
,,	*Norman Court* ..	Shewan	834	197 4	33 0	20 0	Inglis	Baring
,,	*Caliph*	Ritson	914	215 1	36 1	20 4	Hall	Hector
,,	*Osaka*	Lowe	527	165 0	30 1	17 2	Pile	Killick
,,	*Eme*	Sproule	774	199 7	32 6	19 0	Connell	J. Wade
,,	*Doune Castle* ..	Erskine	887	197 1	34 0	18 8	Elder	Skinner
,,	*Ambassador* ..	Duggan	692	176 0	31 3	18 9	Walker, London	W. Lund
1870	*Black Adder* ..	Moore	918	216 6	35 2	20 5	Maudsley, London	J. Willis
,,	*Hallowe'en* ..	Watt	920	216 6	35 2	20 5	Maudsley, London	J. Willis
,,	*Lothair*	Orchard	794	191 8	33 5	19 0	Walker	Killick

APPENDIX B.

Sail Plan of Tea Clipper "Sir Lancelot."

Name of Sail	Gross Yards per Sail-makers' Invoice	Seams Linings, Bands, Bunts, Tablings	Net Yards of Sail Area
Flying jib	250	27	223
Outer jib..	105	18	87
Inner jib	140	19	121
Fore topmast staysail	85	13	72
,, royal	84	15	69
,, topgallantsail	205	50	155
,, upper topsail	230	64	166
,, lower topsail	243	71	172
Foresail	509	96	413
Main topmast staysail	240	22	218
,, staysail	144	15	129
,, middle staysail	105	12	93
,, topgallant staysail	105	12	93
,, royal staysail	95	11	84
,, royal	102	16	86
,, skysail	63	11½	51½
,, topgallantsail	222	53	169
,, upper topsail	236	64	172
,, lower topsail	255	74	181
Mainsail	594	108	486
Mizen staysail	135	17	118
,, topmast staysail	113	16	97
,, topgallant staysail	79	12	67
Crossjack	378	91	287
Mizen topsail	287	55	232
,, topgallantsail	141½	35½	106
,, royal	66½	10½	56
Spanker	266	24	242
Studding Sails—			
Main lower	274	18	256
,, topmast	148	12	136
,, topgallant	73	8	65
,, royal	65	7	58
Fore lower	274	18	256
,, topmast	148	12	136
,, topgallant	73	8	65
,, royal	57	6	51
Running yards, 3 feet by 2 feet	6590	1121½	5468½
= Square feet	39,540	6729	32,811

Extra sails not included:—

Jamie Green	142	yards (about).
Gaff Topsail	105	,, ,,
Ringtail	163	,, ,,
Jib-o'-jib..	99	,, ,,

APPENDIX C.

Spar Measurement of "Normancourt."

BOWSPRIT. Jibboom (extreme length), 68 feet 4½ inches.
Boom end, 2 feet 4½ inches. Flying boom, 12 feet. Outer boom 15 feet.
Inner boom, 17 feet. Heel, 22 feet.

FOREMAST. Topmast, 43 feet (masthead, 7 feet 6 inches).
Topgallant mast, 25 feet.
Royal mast, 15 feet.
Lower mast, 58 feet deck to cap (masthead, 13 feet 3 inches).
Extreme length (deck to truck), 120 feet.
Foreyard (extreme), 71 feet.
Lower topsail yard 61 feet.
Upper topsail yard, 56 feet
Topgallant yard, 41 feet.
Royal yard, 32 feet.

MAINMAST. Top mast, 46 feet 6 inches (masthead, 7 feet 6 inches).
Topgallant mast, 26 feet.
Royal mast, 16 feet
Skysail mast, 10 feet.
Lower mast, 61 feet 6 inches, deck to cap (masthead, 13 feet 6 inches.)
Extreme length (deck to truck), 139 feet.
Main yard, 74 feet
Lower topsail yard, 65 feet.
Upper topsail yard, 60 feet.
Topgallant yard, 44 feet.
Royal yard, 32 feet.
Skysail yard, 24 feet.

MIZENMAST Top mast, 33 feet.
Topgallant mast, 18 feet.
Royal mast, 12 feet.
Lower mast, 50 feet 6 inches, deck to cap (masthead, 10 feet).
Extreme length (deck to truck) 98 feet.
Crossjack yard, 60 feet 6 inches.
Lower topsail yard 50 feet.
Upper topsail yard, 44 feet.
Topgallant yard, 32 feet 4 inches.
Royal yard, 24 feet.
Spanker gaff (extreme), 31 feet; gaff end, 5 feet; span, 21 feet.
Spanker boom, 48 feet.

APPENDIX D.

Log of "Thermopylae" on her Maiden Voyage, 1868-1869.

London to Melbourne.

Date	Lat..	Long.	Dist.	Remarks
	° ′	° ′		
Nov. 5	—	—	—	5.30 p.m., at Gravesend
6	—	—	—	———
7	—	—	—	5 a.m., left Gravesned
8	—	—	—	6 p.m., Lizard N. 20 miles
9	48 30 N.	7 2 W.	168	Var. moderate
10	45 38	13 16	274	S.E., N.W., fresh
11	43 13	15 38	213	Var. moderate
12	41 11	19 24	194	S.S.E. Lost Peter Johnson overboard, ship hove to for an hour without success
13	39 44	22 10	138	S.S.E. strong gales
14	38 40	22 58	69	Var. moderate
15	35 12	21 54	213	North-westerly, strong
16	30 39	22 55	279	North-westerly, fresh
17	29 9	23 43	99	N., S.E., moderate
18	27 38	26 5	200	Sou'-westerly, moderate
19	26 45	24 12	—	Sou'-westerly, light
20	26 32	24 39	—	Sou'-westerly, light
21	25 14	24 32	68	Easterly, light
22	21 39	26 5	228	E., fresh
23	17 18	26 25	268	Nor'-easterly, fresh
24	13 18	25 32	250	E., fresh
25	10 6	24 33	210	Easterly, moderate
26	6 53	23 32	202	South-easterly, moderate
27	4 27	24 3	140	South-easterly, heavy squalls
28	1 23	25 50	228	South-easterly, moderate
29	2 13 S.	29 0	271	South-easterly, fresh
30	6 30	21 8	288	South-easterly, strong
Dec. 1	11 22	31 28	293	Easterly, variable
2	16 14	31 25	294	Easterly, strong
3	20 24	30 26	256	Easterly, moderate
4	23 0	29 0	176	Easterly, light
5	24 32	27 39	118	Easterly, light
6	25 53	27 8	81	Northerly, light
7	27 22	26 28	96	Northerly, light
8	29 4	25 10	123	Northerly, light
9	32 24	22 35	240	Nor'-westerly, fresh gale
10	26 26	18 51	224	Sou'-westerly, blowing a gale
11	38 34	13 2	303	Sou'-westerly, strong
12	39 38	6 34	314	W.S.W., strong
13	40 34	0 25 E.	324	S.W., strong
14	40 51	6 33	280	Var., moderate
15	41 51	11 19	230	Northerly, fresh
16	42 29	17 30	282	Nor'-westerly, moderate
17	43 6	23 41	278	Nor'-westerly, strong
18	43 9	28 29	211	Nor'-easterly, fresh
19	43 44	34 56	284	Nor'-easterly, strong
20	43 57	40 30	240	Northerly gale
21	43 35	47 34	305	Northerly gale
22	43 45	54 18	290	Northerly gale
23	42 57	61 17	310	Northerly gale
24	43 6	67 51	266	Northerly, strong

Log of "Thermopylae," 1868-1869—*Continued.*

London to Melbourne.

Date	Lat.	Long.	Dist.	Remarks
	° ′	° ′		
Dec. 25	42 57	74 26	312	Northerly, strong
26	43 22	80 28	265	Northerly, fresh
27	43 15	85 41	229	Northerly, fresh
28	43 22	90 40	222	Easterly, fresh
29	43 40	94 55	185	N.E., light
30	43 11	102 11	320	S.W., gale
31	43 4	106 43	200	N.N.W., moderate
1869				
Jan. 1	43 10	111 54	228	N.N.W., moderate
2	43 7	117 14	248	N.N.W., fresh
3	42 7	124 36	330	Northerly, strong
4	40 39	131 18	326	Northerly, strong
5	39 48	136 14	225	Sou'-westerly, moderate
6	38 41	140 18	202	S.E., Percy Island
7				Cape Otway, N. ½° W., 12 miles
8				Calm and light
9				7 p.m., came to anchor in Port Phillip Harb.
		Newcastle, N.S.W., to Shanghai.		
Feb. 10	—	—	—	Left the harbour, 7.30 a.m.
11	—	—	60	E.N.E. to S.E., calm
12	32 46 S.	156 3 E.	152	N.E. and E., very unsteady
13	32 13	158 26	125	N.E. and N., passed Lord Howe's Island
14	28 30	160 55	250	N., strong, squally
15	23 32	162 16	300	N., clear
16	19 47	161 58	230	N.W., clear
17	15 36	162 11	251	N. ½° E., heavy, squally
18	13 31	163 24	145	N.N.E., thunder and lightning
19	12 16	163 17	75	N., heavy rain
20	8 35	164 0	224	N., heavy rain and thunder
21	4 16	165 24	262	N. and E., lightning
22	1 35	166 48	180	N.N.E., heavy squalls
23	0 19	166 50	75	N.W. by W., off Pleasant Head, got quantity of jugs and cocoanuts
24	1 14 N.	165 5	130	N.W.
25	3 24	162 25	200	N.W. ½° W., squally
26	6 47	159 58	250	N.W. by N., clear
27	10 28	156 35	297	N.W. by W., fresh
28	13 28	152 4	298	N.W. by W., fresh
Mar. 1	15 54	148 25	256	N.W. by W., squally
2	17 14	146 2	160	N.W. by W., passed between Faraltan and Guguants
3	19 56	142 35	255	N.W. by W. ½° W., squally
4	21 46	139 48	200	N.W. by W. ½° W., light
5	22 23	138 19	82	N.W. by W., light
6	23 32	136 43	110	N.W. by W., light and variable
7	25 23	133 34	202	N.N.W., passed *Golden*, Sydney to Shanghai, 59 days
8	26 57	129 26	239	W.N.W., passed Fok Island
9	29 30	126 11	230	W. by N., heavy squalls, thunder & lightning
10	31 20	124 0	200	W.N.W. Off Vido. Got pilot. Passage pilot to pilot 28 days, quickest on record
13	—	—	—	Shanghai. Thick and calm.

Log of "Thermopylae," 1868-1869—*Continued.*

Foochow to London.

Date	Lat	Long.	Dist.	Remarks
	° '	° '		
July 3	—	—	—	5 a.m., proceeded down in tow
4	—	—	—	Pinnacle Island, W by N.
5	—	—	—	3 p.m., passed Adam's Point
6	23 6 N.	126 32 E.	174	Sou'-westerly, fresh
7	23 2	126 1	48	Var., moderate
8	21 13	123 59	157	South-easterly, squally
9	19 13	120 28	233	South-easterly, fresh
10	18 36	118 37	115	South-westerly, light
11	18 16	116 2	148	South-easterly, fresh
12	17 37	112 38	200	Southerly, fresh
13	16 43	109 49	176	Southerly, fresh
14	15 43	109 5	63	South-easterly, light
15	14 6	110 7	114	South-westerly, light
16	12 37	109 29	96	South-westerly, light. Cape Varella 6 miles
17	11 47	109 30	50	South-westerly, light
18	8 16	109 49	212	Westerly, strong and squally
19	5 9	109 21	191	South-westerly, strong and squally
20	4 16	109 17	53	South-westerly, moderate
21	3 28	109 22	48	South-westerly, light
22	2 40	109 54	58	South-westerly, light
23	2 6	109 6	61	Var., light
24	0 51	108 40	77	Boorand Island, E. by S, 10 miles
25	0 45 S.	108 34	96	South-easterly, light
26	1 28	107 48	63	Var., light
27	3 15	106 59	116	Spoke *Achilles*, 10 days out from Foochow
28	—	—	165	6 a.m., Anjer Light, S S.W., 8 miles
29	7 54	101 56	223	South-easterly, fresh
30	9 22	97 21	284	South-easterly, squally
31	10 59	93 10	267	South-easterly, fresh
Aug. 1	12 42	88 43	290	S.S.E., strong
2	14 31	83 28	318	E.S.E, strong
3	16 5	79 44	236	South-easterly, moderate. Spoke *Leander*
4	17 30	76 33	203	S.E, moderate. *Leander* 10 miles
5	18 45	72 58	217	S.E., fresh. *Leander* 14 miles
6	19 16	71 26	97	S.E., light
7	19 4	68 28	170	S.W to S.E. Heavy gale, and sea washed away headrail
8	21 11	63 53	249	S. by E., under topsails and courses
9	23 4	59 0	295	S. by E., strong
10	24 30	54 55	246	S. by E., all plain sail
11	26 9	51 23	216	S. by E., var., plain sail & port studding sails
12	27 25	48 30	185	E.N.E., moderate
13	29 7	45 24	192	E, light
14	—	—	170	W.S.W., var.
15	30 23	38 29	200	S., strong gale with squalls
16	31 20	35 0	198	E., light
17	34 20	33 35	110	S.E., steamer astern like *Achilles*; sunset, breeze increasing, leaving her out of sight
18	34 2	29 39	270	N.N.E., fresh
19	35 6	24 0	240	S.W. by S., fresh, strong current to S.W.
20	35 8	20 4	196	N.E., fog and calm at noon
21	34 45	18 10	100	W.S.W., increasing, rounded Cape of Good Hope, heavy sea
22	31 53	13 26	302	S., all plain sail set
23	29 9	9 29	262	S.S.E., all plain sail set and studding sails

Log of "Thermopylae," 1868-1869—*Continued.*

FOOCHOW TO LONDON.

Date	Lat.	Long.	Dist.	Remarks.
Aug. 24	26° 14′	5° 19′	284	S.E., all possible sail
25	23 13	1 50	264	S.E. by S., all possible sail
26	20 44	0 53	212	N.E., and backing to S.E.
27	19 9	2 49 W.	146	S.E., light
28	17 29	4 58	158	S.E., light
29	15 36	7 33	187	S.E., light
30	13 19	10 5	201	S.E., light
31	11 16	12 16	190	S.E., light
Sept. 1	9 6	14 8	164	S.E. by S., light
2	7 11	16 0	158	S.E. by S., light
3	5 9	18 2	172	S.E. by S., light
4	3 19	19 51	156	S.E. by S., light
5	1 10	21 46	172	S.E. by S., light, strong current to W.
6	0 55 N.	23 4	146	S.E., light
7	2 57	25 4	174	S.E. by S., light
8	5 51	26 7	184	S.S.W., fresh
9	10 0	27 6	257	S.W., very squally
10	12 16	27 16	140	S., light var.
11	13 10	27 0	60	N.N.E., var., squally
12	16 33	30 9	273	N.E., trade winds
13	20 5	32 58	270	N.E., trade winds
14	24 0	35 23	272	N.E. by E., trade winds
15	26 45	36 15	172	E. by N., light
16	27 39	36 18	54	E. by N., light and calm
17	28 0	36 23	21	E. by N., light and airy
18	28 56	36 5	58	S., light
19	30 18	35 45	52	S., light
20	32 37	35 5	144	S., light
21	33 45	34 18	85	W., light rain
22	36 4	34 4	140	W. light breeze
23	39 18	33 30	200	S.W., squally, rain
24	42 37	30 17	245	W., squally, heavy sea
25	44 10	26 16	200	W., light and variable
26	45 14	22 59	158	S.W. to N.W.
27	46 8	18 34	200	W., bar falling rapidly
28	47 15	14 0	202	S.W., bar falling rapidly, very low
29	48 30	9 13	200	S.W., bar falling rapidly very low
30	—	—	200	S. by E., noon Lizard, N., 8 miles
Oct. 1	—	—	—	Beachy Head, E., 20 miles at noon. 5 p.m., Dungeness, got pilot.

APPENDIX E.

Abstract Log of "Hallowe'en," Captain James Watt.

SHANGHAI TO LONDON.

Date	Lat.	Long.	Course	Dist.	Remarks.
1873	° ′	° ′	°		
Nov 19	—	—	—	—	1 p.m., passed the lightship, fresh northerly breeze. 3.30 p.m., off North Saddle
20	28 12 N.	121 50 E.	—	210	Mod. northerly breeze P.M., light winds
21	26 35	120 51	S. 28 W.	110	North-easterly, light. P.M., increasing
22	23 32	118 40	S. 35 W.	215	North-easterly, smart steady breezes
23	19 18	117 18	S. 17 W.	266	Fresh N.E. monsoon
24	16 17	114 57	S. 36 W.	225	North-easterly, steady and fine
25	13 43	112 55	S. 37 W.	193	North-easterly, moderate and fine
26	11 10	111 11	S. 35 W.	176	North-easterly, moderate and fine
27	8 28	109 12	S. 33 W.	192	North-easterly, moderate and fine
28	6 29	107 31	S. 40 W.	120	Variable, E. to S.E. unsteady, squally
29	4 28	106 49	S. 19 W.	121	Easterly, light and showery
30	2 0	106 57	S. 3 E.	148	Var., northerly to S.E.
Dec. 1	0 50 N.	106 57	South	70	Var. and calms, thick rain
2	1 26 S.	107 26	S. 13 E.	134	Fresh westerly & North-westerly. P.M. mod.
3	3 25	106 50	S. 17 W.	124	N.W. to S.W. 2.30 a.m., anchored off ent. Macclesfield Channel. 9 a.m., got underweigh.
4	5 9	106 4	S. 24 W.	114	Var., westerly and squally. 5.30 p.m., anchored off North Isl.
5	6 10	105 40	S. 22 W.	66	3 a.m., got underweigh. Wind mod. W.N.W. to W.S.W.
6	6 47	104 58	S. 48 W.	56	Var., W.N.W. to W.S.W. Noon, 15 miles W. of Java Head.
7	8 46	104 34	S. 11 W.	121	Unsteady southerly wind.
8	9 42	100 25	S. 77 W.	251	Smart steady breezes, S. by E.
9	11 3	98 8	S. 59 W.	158	S.S.E., mod. and fine
10	12 15	94 53	S. 69 W.	205	S.S.E., mod. & fine. P.M., light
11	13 19	92 33	S. 65 W.	151	South-easterly, light winds. P.M., unsteady
12	14 17	90 35	S. 63 W.	129	S.S.E. to E., light winds
13	15 55	87 33	S. 61 W.	201	S.E. by S., fine steady breezes
14	17 38	84 5	S. 63 W.	225	South-easterly, fine steady breezes
15	19 13	81 0	S. 62 W.	200	East. Mod. trade and fine
16	20 25	77 15	S. 71 W.	224	East. Mod. trade and fine
17	21 30	74 15	S. 69 W.	180	E. to S.E. Mod. and unsteady
18	22 21	71 35	S. 71 W.	157	E. to E.N.E., light winds
19	23 8	69 13	S. 70 W.	139	E.N.E. to N.E. and N.N.E., light winds. Heavy S.W. swell.
20	24 0 S.	66 30 E.	S. 71 W.	157	N.W. to N.N.W. fresh. 11 a.m., shift to S.W. P.M., blowing hard
21	24 38	62 16	S. 81 W.	235	Southerly, strong and squally. Heavy head sea
22	25 43	58 21	S. 73 W.	223	S. to S.E., strong steady breeze. P.M., moderate and unsteady
23	26 27	55 51	S. 72 W.	142	Easterly to E.N.E. Light winds and fine
24	27 13	53 41	S. 69 W.	126	E.N.E. to N.E. Light winds and fine
25	28 24	49 49	S. 71 W.	217	N.E. smart breeze

Abstract Log of "Hallowe'en"—cont.

Shanghai to London.

Date	Lat.	Long.	Course	Dist.	Remarks
1873	° ′	° ′	°		
Dec. 26	29 21	46 50	S. 70 W.	167	N.E. mod. 4 a m. to 11, light to calm. 11 a.m., fresh S.S.E.
27	30 29	40 53	S. 78 W.	326	S.S.E. to E.S.E. strong and squally, nasty cross sea
28	31 58	35 52	S. 71 W.	273	Easterly fresh breeze and smooth water
29	32 54	32 32	S. 71 W.	179	Easterly to N.E. by E. Mod. breeze. P.M., light and fine
30	33 24	29 24	S. 79 W.	162	N.E. to S.W. 4 a.m., shift of wind, light, but heavy head sea
31	34 32	26 30	S. 65 W.	159	Var., southerly and westerly, mod. and unsteady. 3 p.m., commenced to blow. 8 p.m., fresh gale
1874					
Jan. 1	35 32	25 35	S. 37 W.	75	North-westerly. Blowing hard, gusty heavy confused sea
2	34 56	24 24	N. 58 W.	69	North-westerly. Fresh breeze and heavy confused sea
3	34 54	23 9	N. 88 W.	60	N.W. to S.E., light. 7 a.m., increasing heavy S.W. sea
4	35 17	19 48	S. 82 W.	166	S.E., mod. and hazy, heavy head sea. P.M., 8-12, strong breeze
5	32 38	15 13	N. 55 W.	279	S.E. to S., strong, all sail and all studding sails set
6	29 38	11 19	N. 48 W.	270	South to S.S.E., mod., smooth water. Going 13-14 knots
7	27 32	8 21	N. 51 W.	202	South to S.S.E., mod. and fine
8	25 59	6 30	N. 47 W.	136	Variable, S.S.E. to S.W. Light winds and fine
9	24 6	4 24	N. 45 W	160	S.S.E., light winds and fine. P.M., moderate
10	22 6	2 2 E.	N. 47 W.	178	S.S.E., moderate winds and fine
11	20 12	0 9 W	N. 47 W.	167	S.S.E., moderate winds and fine
12	18 6	2 44	N. 49 W.	194	S.S.E., fine steady trades
13	16 6	5 10	N. 49 W.	185	S.S.E., moderate trades. p.m., signalled St. Helena
14	14 34	7 20	N. 54 W.	154	S.S.E., moderate trades
15	12 28 S.	9 23 W.	N. 43 W.	173	S.E. by S. to S.E. mod. and fine
16	10 33	11 9	N. 42 W.	155	S.S.E., moderate and fine
17	8 35	12 46	N. 39 W.	152	S.E. by S. to S. by E., mod. and fine
18	6 33	14 38	N. 42 W.	166	S.E. by S. to S. by E., mod. and fine
19	4 28	16 38	N. 44 W.	165	S.S.E., moderate and fine
20	2 38	18 41	N. 48 W.	165	S.S.E., mod. northerly swell
21	0 18 S.	20 40	N. 40 W.	185	S. by E. and S.S.E. Fine steady breezes
22	2 26 N.	21 14	N. 12 W.	168	South to S.E., unsteady. Heavy northerly swell
23	4 10	21 16	N. 1 W.	104	S.E. light, unsteady. P.M., light airs and sultry
24	5 13	21 18	N. 2 W.	63	Easterly & north-easterly, light airs find calms
25	6 20	22 34	N. 48 W.	102	N.N.E. to N.E., moderate
26	8 20	25 13	N. 53 W.	199	N.E. by N. to N.E., smart steady breezes
27	11 11	27 24	N. 37 W.	214	N.E. to N.E. by E. Fine steady trades.

Abstract Log of "Hallowe'en"—cont.

SHANGHAI TO LONDON.

Date	Lat.	Long.	Course	Dist.	Remarks
1874	° ′	° ′	°		
Jan. 28	14 00	29 48	N. 40 W.	220	N.E. to N.E. by E. Fine steady trades
29	16 36	31 56	N. 38 W.	198	N.E. to E.N.E. and N.N.E., light steady breeze
30	19 3	33 18	N. 28 W.	166	Var. N. by E. to east, very unsettled winds in force and direction
31	21 2	33 45	N. 13 W.	122	Easterly. Light variable winds
Feb. 1	22 44	34 6	N. 11 W.	104	Easterly to south. Calms and light variable winds
2	23 46	34 31	N. 20 W.	66	Southerly, east to N.E. Calms and variable. Heavy rain
3	24 59	35 45	N. 43 W.	100	N.E. to N.N.E., calms and light unsteady P.M., mod.
4	26 28	34 44	N. 32 E.	105	N. to N.N.W. and N.N.E., unsteady, moderate. P.M., fresh and squally
5	27 16	32 2	N. 12 E.	152	N.N.E., N.E. by N., fresh. P.M., light and heavy N.W. swell
6	27 36	31 44	N. 39 E.	26	Var.east to south, light airs. P.M., increasing
7	30 30	30 55	N. 14 E.	179	S.W. to S.S.W., fine steady breeze. P.M., moderate and heavy N.W. swell.
8	32 28	29 50	N. 25 E.	130	N.W. to west and S.W., light and unsteady. P.M., mod.
9	35 25	27 47	N. 30 E.	204	S.W. to W.S.W. Fresh breeze and thick.
10	38 55 N.	26 19 W	N. 18 E.	221	Westerly and N.W. Fresh breeze and squally. P.M., strong
11	41 48	22 5	N. 48 E.	262	N.W. west. A.M., blowing strong and squally P.M., moderate
12	44 4	18 27	N. 49 E.	210	West to S.W. to N.W. Mod. to fresh increasing
13	45 47	14 6	N. 61 E.	211	N.W. to west and W.S.W. Fresh breezes and hazy
14	48 21	8 53	N. 54 E.	264	W.S.W. to W. by S. Strong breeze and squally
15	50 9	3 38	N. 62 E.	233	W.S.W. to S.W. Fresh and squally
16	—	—	—	—	At 8.30 a.m., got the pilot on board. 11.30 a.m., passed through the Downs. Noon, off the North Foreland. 1.30 p.m., taken in tow near the Tongue Lightship. Blowing hard from S.W. 2 p.m., wind west. 8 p.m., brought up at the Chapman
17	—	—	—	—	Arrived at Gravesend. Blowing hard from S.W., and thick rain

APPENDIX F.

Complete List of "Thermopylae's" Outward and Homeward Passages under the Aberdeen White Star House Flag, 1868-1890

Year	Captain	Left	On	Arrived	On	Days Out
1868-69	Kemball	London	Nov. 7	Melbourne	Jan. 9	63
1869	,,	Foochow	July 3	London	Oct. 1	90
1869-70	,,	London	Nov. 10	Melbourne	Jan. 25	76
1870	,,	Foochow	July 29	London	Nov. 12	106
1870-71	,,	London	Dec. 27	Melbourne	Mar. 2	65
1871	,,	Shanghai	June 22	London	Oct. 6	106
1871-72	,,	London	Nov. 11	Melbourne	Jan. 23	73
1872	,,	Shanghai	June 18	London	Oct. 11	115
1872-73	,,	London	Nov. 13	Melbourne	Jan. 27	75
1873	,,	Shanghai	July 11	London	Oct. 20	101
1873-74	,,	London	Dec. 2	Melbourne	Feb. 17	77
1874	,,	Shanghai	July 18	London	Oct. 27	101
1874-75	Matheson	London	Nov. 28	Melbourne	Feb. 6	70
1875	,,	Foochow	July 8	London	Oct. 31	115
1875-76	,,	London	Nov. 29	Melbourne	Feb. 9	72
1876	,,	Foochow	July 29	Scillies	Nov. 21	115
1876-77	,,	London	Dec. 19	Melbourne	Mar. 14	85
1877	,,	Shanghai	July 8	London	Oct. 20	104
1877-78	,,	London	Dec. 3	Melbourne	Feb. 17	76
1878-79	,,	Shanghai	Nov. 27	London	Mar. 17	110
1879	,,	London	June 2	Sydney	Sept. 3	93
1879-80	,,	Sydney	Nov. 18	London	Feb. 7	81
1880	,,	London	May 21	Sydney	Aug. 11	82
1880-81	,,	Sydney	Oct. 14	London	Jan. 12	88
1881	Henderson	London	Mar. 10	Sydney	June 6	90
1881-82	,,	Foochow	Oct. 30	London	Feb. 15	107
1882	,,	Lizard Light	Mar. 21	Sydney	June 2	73
1882	,,	Sydney	Oct. 14	London	Dec. 29	77
1883	,,	London	Jan. 21	Sydney	May 9	109
1883-84	,,	Sydney	Oct. 31	London	Jan. 26	87
1884	Allan	London	Feb. 25	Sydney	May 18	82
1884	,,	Sydney	Oct. 6	Prawle Point	Dec. 23	78
1885	,,	London	Jan. 19	Melbourne	April 8	79
1885-86	,,	Sydney	Oct. 18	London	Jan. 6	80
1886	,,	London	Feb. 16	Sydney	May 20	93
1886-87	,,	Sydney	Oct. 24	London	Jan. 19	87
1887	,,	London	May 11	Sydney	July 25	75
1887-88	,,	Sydney	Oct. 16	London	Jan. 3	79
1888	Jenkins	London	Feb. 16	Sydney	May 7	80
1888	,,	Sydney	June 9	The Lizard*	Sept. 22	105
1888-89	,,	London	Oct. 30	Sydney	Jan. 29	91
1889	,,	Sydney	Mar. 26	London	June 29	95
1889	,,	London	Aug. 10	Sydney	Nov. 1	83
1890	,,	Sydney	Jan. 9	Deal*	April 8	89

*For Rotterdam.

APPENDIX G.

Complete List of "Cutty Sark's" Australian Passages.

Year	Captain	Left	On	Arrived	On	Days Out
1872-3	Moore	Off Start	Dec. 4	Melbourne	Feb. 11	69
1873-4	Tiptaft	Off Portland	Dec. 15	Sydney	Mar. 4	79
1874-5	,,	Off Start	Nov. 21	,,	Feb. 2	73
*1875-6	,,	Off Portland	Nov. 27	,,	Feb. 12	77
1876-7	,,	Off Start	Oct. 23	,,	Jan. 10	79
1877-8	,,	Off Lizard	Dec. 6	,,	Feb. 16	72
1883	Moore	Channel	July 24	Newcastle	Oct. 10	79
1883-4	,,	Newcastle	Dec. 28	Off Beachy Head	Mar. 20	82
1884	,,	Channel	June 18	Newcastle	Sept. 5	79
1884-5	,,	Newcastle	Dec. 9	London	Feb. 27	80
1885	Woodget	Off Start	April 3	Sydney	June 19	77
1885	,,	Sydney	Oct. 16	London	Dec. 27	72
1887	,,	Sydney	Mar. 26	,,	June 6	72
1887	,,	Channel	Aug. 19 dismasted	Newcastle	Nov. 15	88
1887-8	,,	Newcastle	Dec. 28	Dungeness	Mar. 8	71
1888	,,	London	May 20	Sydney	Aug. 5	77
1889-9	,,	Sydney	Oct. 26	Start	Jan. 18	84
1889	,,	London	May 8	Sydney	July 26	79
1889-90	,,	Sydney	Nov. 3	London	Jan. 16	74
1890	,,	Off the Lizard	May 21	Sydney	Aug. 4	75
1890-1	,,	Sydney	Dec. 13	London	Mar. 17	94
1891	,,	London	April 25	Sydney	July 14	80
1891-2	,,	Sydney	Nov. 5	London	Jan. 28	84
1892	,,	London	Aug. 12	Newcastle N.S.W.	Nov. 9	89
1893	,,	Sydney	Jan. 7	Off St. Catherines	April 11	94
1893	,,	Off the Lizard	Aug. 8	Sydney	Oct. 29	82
1893-4	,,	Sydney	Dec. 24	Hull	Mar. 28	94
1894	,,	Off Portland	June 27	Brisbane	Sept. 15	80
1894-5	,,	Off Brisbane	Dec. 30	London	Mar. 26	86

Cutty Sark was 50 miles south of Melbourne on 64th day out, but owing to strong headwinds had to go round Tasmania.

Running easting down she did 2163 miles in 6 days.

APPENDIX II.

Abstract Log of "Ariel" Captain Keay, Foochow to London, 1866.
(*From Captain Keay's private Journal.*)

Tuesday, 29th May, 1866.—5 a.m., hove up and 5.30 proceeded, towing alongside down the river. 8.30, nearing Sharp Rock; discharged China pilot. 9, tried to get steamer ahead to tow but very soon she sheered wide to port, could not recover command of the helm and obliged us to anchor. Again tried to tow alongside and proceeded outside the wreck of *Childers*, but were damaging steamer's sponsons and our side so much that we had to cast off and pilot would not risk going on as steamer could not be relied upon to get ahead in time, tide already having fallen, therefore anchored in hopes of getting on to-night. The *Fiery Cross* towed past us and went to sea all safe drawing less water. We are now forward 18 feet 8 inches and aft 18 feet 3 inches, out of trim but hope to have her right soon. 8 to 10 p.m., had the night been clear would have gone to sea but showery thick weather, pilot would not venture. Wind N.E. moderate.

Wednesday, 30th May, 1866.—5 a.m., turned to and brought aft to abreast mizen mast 30 fathom of each bower chain and the stream chain, also gin chains and 12 casks of salt provisions; lashed alongside the after-boats; stowed all the studding-sail gear between said boats in the gratings and passed the holystones aft out of lower forecastle to trim ship more by the stern. 8.30 a.m., hove short and got the steamer ahead, towropes fast one from each bow to his quarters. 9 a.m., weighed and proceeded under tow; the *Taeping* and *Serica* following us. 10.30 a.m., were well outside the Outer Knoll, cast off the tug and hove to for his boat to fetch away our pilot, Smidt. They lowered the boat, the steamer going ahead, she filled, men were saved but they were so long picking up the men and boat that we signalled for a pilot boat to take away our pilot. At 11.10 a.m., filled the mainyard and steered S. by E. ½ E. for Turnabout Island. Made sail and set fore-topmast and lower stunsail and skysail. Rain and moderate N.E. wind. We left *Taeping* and *Serica* a little. Noon, S.E. point of White Dogs E.N.E. about 6 miles. 1 p.m., Warning Rocks W. by S. 3 miles. 3 p.m., South Point of Turnabout Island N.W. 2 miles. Set fore-top-gallant stunsail. Same wind and weather.

Thursday, 31st May, 1866.—Same wind and weather. Noon, S.E. of Brother. bore N.W. by N. 6 miles. Lat. 23° 27′ N., long. 117° 45′ E. Distance 190 miles. 3 p.m., saw Lamock Islands on starboard quarter. Watch putting on chafing gear.

Friday, 1st June, 1866.—Cloudy with showers and moderate N.E. wind. A.M., stowed 23 half-chests and 16 chests of tea in after cabin in locker-heads. Noon, lat. 21° 10′ N., long. 115° 9′ E. Distance 195 miles. Experienced no current against us. P.M., carpenter secured tea in cabin with stanchions, etc. Took three doors off their hinges to facilitate stowing hawsers and lines, etc., in my cabin and in starboard passage to get ship more in trim. 6.30 p.m., altered course more southerly to bring wind two points on the quarter for speed.

Saturday, 2nd June, 1866.—Wind moderate from N.E. and clear. 5 a.m., wind from N.N.W., altered course to S.W. and brought wind on port quarter, partly for speed and determined to go west of Paracels. *Taeping* in sight to S.E. by E. Watch employed putting 30 fathoms of each cable below in sails of cabin and first coiled the Europe hawsers in my cabin, also put nine casks of pork in after store-room and lashed seven of beef on main hatch, only leaving two of beef in lower forecastle. Ship seems to steer very easy and is probably almost in trim. Noon, lat. 17° 51′ N., long. 112° 57′ E. Distance 240 miles. P.M., wind getting very light and veering between N, N.E., and S.E. Setting and taking in stunsails and trimming sails as required.

Sunday, 3rd June, 1866.—Light airs and calm, sky overcast at times, steering along north of Paracels. Noon, lat. 17° 14′ N., 111° 32′ E. Distance 83 miles. 1 p.m., from aloft saw north shoal of Paracels bearing south about 8 miles distant appearing in long patches of breakers with black rocks showing a little above water, very faint baffling airs, steered west to get clear of the shoal. 4 p.m., still bearing between S. ½° W. and S.S.E.

Monday, 4th June, 1866.—Watch putting on chafing gear, drying awnings, towing warps, etc. Vyse making the stunsail bonnets. Carpenter making a main topgallant mast; first-rate spar bought from Robinson of Pagoda Anchorage. Found ship was fully 6 inches by the stern, therefore shifted spars forward to their proper place to make the decks more clear. This seemed to alter the trim again so that ship is about 4 inches by the stern, viz., 18 feet 5 inches aft and 18 feet 1 inch forward. Noon lat. 16° 55′ N., long. 110° 21′ E. Distance 172 miles. P.M., calm and baffling airs. 5 p.m., light wind from S.E.

Tuesday, 5th June, 1866.—Same wind and clear with much lightning on western horizon. 9 a.m., Pulo Canton in sight W. ½° S. 15 miles. Hawsers again up to dry, having been thoroughly soaked. Sent down royal stunsail booms and gear to be snug aloft for beating. Watch lacing the foot of upper topsails to the jackstays of lower topsail yards. Scraped and oiled the bower anchors and chains outside the house to lessen rust; opened the quarter-hatch to ventilate the hold better while weather is fine. Noon, lat. 15° 18′ W., long. 109° 23′ E. Distance 117 miles. P.M., light wind from S. by E. to S.S.W.; tacking as necessary; keeping inshore in hope of land breeze at night. Midnight, fresh breeze came off from S.W., tacked to S.S.E. Land of Cochin China about 4 miles distant.

Wednesday, 6th June, 1866.—Light southerly winds, almost calm occasionally. Painted the rails over brass work to save polishing on the passage; also painted outside where bare with chafe of fenders and cargo boats, etc. Putting on and refitting chafing gear. Carpenter finished topgallant-mast, coated it thick with pine oil; is making a topmast studding sailyard for enlarged topmast stunsail. Noon lat. 13° 24′ N., long. 109° 58′ E. Distance 122 miles. Light airs from southward, stood in towards shore in hopes of land wind; tried to tack, too little way, wore round to S.E., faint airs from S. to S.W. and calms; not steering. Midnight, the same weather.

Thursday, 7th June, 1866.—2 a.m., a faint air from S.W.; began to steer; current favourable. Noon, lat. 12° 21′ N., long. 109° 28′ E. Distance 72 miles. Try the pumps every evening about 7. Get very little water out. 4 a.m., moderate S.S.W. wind, tacked to westward. 7.45 a.m., about 8 miles off shore near Fisher's Island. Wind came more off the land; tacked to S. by E. 11 p.m., a small-vessel ran nearly into us. Midnight, wind in strong gusts; took in small sails.

Friday, 8th June, 1866.—Moderate and steady S.W. monsoon; tolerably clear. Watch employed about rigging; carpenter making bowline bulls' eyes. Noon, lat. 9° 55′ N., long. 110° 4′ E. Distance 150 miles. Made 1½ points leeward off course and 12′ less southing than by dead reckoning. Set to leeward say 1½ knots 36 miles N.E. per day from 8 p.m. of 7th to noon of 8th; steered to pass east of Prince of Wales bank. 6 to 8 a.m., passed two ships running. 5 to 10 p.m., showery and wind from S.S.W. to W.S.W. moderate.

Saturday, 9th June, 1866.—Same wind. No bottom at 16 fathoms at 5 a.m. Noon, lat. 7° 22′ N., long. 110° 27′ E. Distance 155 miles. Found there had been no current. Dead reckoning and observations agree to a mile; hence passed some 10 miles to windward instead of to leeward of aforesaid bank, probably the N.E. current is not wide. Watch variously employed about rigging. Carpenter rough-making two topmast studdingsail booms of two Foochow pines, an inch or more in diameter. 5 p.m., signalled the *Taeping* about 3 miles to leeward standing west. 6 p.m., she tacked and followed in our wake. Light S.W. wind and clear.

Sunday, 10th June, 1866.—Same weather. 6 a.m. *Taeping* about 4 miles on our lee quarter. Signalised that they had passed the *Fiery Cross* on Friday. We are thus in all probability the headmost ship so far. Noon, lat. 5° 14′ N., long. 111° 20′ E. Distance 142 miles. P.M., light wind from S.W. by S. and clear warm weather; no current.

Monday, 11th June, 1866.—Employed stripping and reserving service of lower and topmast rigging where needed. Noon, lat. 4° 11′ N., long. 111° 47′ E. Distance 69 miles. P.M., very faint airs baffling about; trimming yards, etc. Midnight, light air from westward.

Tuesday, 12th June, 1866.—1.45 a.m., a cloud rose rapidly from W.N.W., strong breeze and little or no rain. In all staysails; clewed up royals and down flying jib. 2 a.m., moderate; set all sail again. Sky clearing and wind gradually hauling after 4 a.m. to S.W. and S. 6.30 a.m., rain and moderate S.S.E. wind; tacked to S.W. Rain gradually thinned off and at noon clear weather and light S.E. wind. port stunsails set forward. Lat. 3° 16′ N., long. 11° 20′ E. Distance 54 miles. A three-masted schooner on starboard bow. 4 p.m., abeam. 5 p.m., calm and faint airs from N.E.; clear warm weather. Carpenter making a box of 3-inch deals 12 feet by 3½ by 2 to hold spare kedges, anchor stocks, yard-hoops, etc., and fill it up with coal from coal locker in forepeak to lighten her there, as ship is only 2 inches by the stern since the spars were put in their places again. Present draft 18 feet 3 inches aft and 18 feet 1 inch forward. Will use said box also to trim to windward.

Wednesday, 13th June, 1866.—Calms and baffling airs. Watch trimming yards, setting and taking in stunsails, etc. Noon, lat. 3° 9′ N., long. 111° 6′ E. Distance 23 miles. P.M., same schooner in company. Light variable S.W. and W. winds backing to S.W. and becoming squally towards midnight. Midnight, tacked to N.W.

Thursday. 14th June, 1866.—Wind veering with very threatening sky, squalls, lightning and rain. All small sails in, and fore and mizen topgallant sails, also main for a short time. Towards daylight more settled wind S.W. with showers and squalls. Set fore and mizen topgallant sails. Main royal and flying jib set and taken in as necessary. Noon, lat. 2° 16′ N., long. (by account) 110° 42′ E. Distance 66 miles. P.M., clearing gradually. Po Point and Tanjong Sipang on port beam and Cape Datu about S.W. by W. Moderate and wind hauling to S.S.W. and S. off the land. Set all small sails and fore-topmast and topgallant stunsails. Midnight, Cape Datu bore south about 5 miles distant. Light breeze from S.S.E. suddenly failed, weather clear.

Friday, 15th June, 1866.—1 a.m., faint air from westward. Got trimmed on the port tack and steering about 2 a.m. 3 a.m., nearly calm. 3.30 a.m., faint air from S.S.E.; S.S.E. airs, calms and baffling till 2 p.m., Noon lat. 2° 4′ N., long. 109° 22′ E. Distance 80 miles. P.M., clear weather, smooth sea, light breeze springing up from W.S.W. 4 p.m., tacked to southward. Marundum bearing west 5 or 6 miles. 6 p.m., tacked to W.N.W. 9 p.m., tacked to S.S.E. 11 p.m., caught aback with moderate S.E. wind, braced round on port tack. Midnight, all sail set; heading S.W. by S.; clear weather.

Saturday, 16th June, 1866.—Same weather. Watch employed about the rigging. Carpenter making a grating for forepart of galley door. Had the box finished and stowed small anchors, etc., in it and over 1 ton of coal lashed and chocked before the quarter hatch. Noon, Tambelan in sight west; tacked to E by S., wind S.S.E., moderate and clear. Lat. 1° 1′ N., long. 108° 1′ E. Distance 102 miles. P.M., Boerang Islands in sight to S.E. Had life-buoys covered with duck and twelve brass hooped buckets also, fancy plate on handles and knots, and painted handles and inside of buckets, outside kept bright, 7, tacked off-shore. 9, tacked in-shore. 11, got wind off the land, very light, tacked to southward.

Sunday, 17th June, 1866.—Fine clear weather and south-easterly wind, moderate and light. 9 a.m., Datu Island N.E. ¾° N. about 20 miles. Noon, lat. 0° 28′ S., long. 108° 15′ E. Distance 101 miles. 6 p.m., tacked to eastward. 11 p.m., tacked to S.S.W.

Monday, 18th June, 1866.—Same wind and weather. 9 a.m., Billiter Island in sight ahead. 11 a.m., abreast of N.W. Island 1½ miles distant. Dutch ship *Willein* in company, bound to Batavia, promised to report us at Lloyd's. Passed her 2 miles to her 1. Noon, lat. 2° 27′ S., long. 107° 29′ E. Distance 133 miles. From 3.30 p.m. to 7 p.m., beat through east of north and Table Island, then steered to south-westward between Vansittart Shoal and Embelton Island and shoals. 10.15 p.m., had passed to leeward of Fairlee Rock, probably 6 miles distant. Hauled up S. by W., wind moderate from south-eastward.

Tuesday, 19th June, 1866.—Same wind and fine clear weather. Numbers of ships seen these three days past. 11.30 a.m., passed close to westward of North Watcher Island and wind fell very light. Noon, lat. 5° 15′ S., long. 106° 36′ E. Distance 164 miles. P.M. faint airs from eastward. Watch still at work stripping and reserving lower and topmast rigging. Midnight, light wind from E. to S.E. and clear.

Wednesday, 20th June 1866.—1 a.m., approaching Cape St. Nicholas. 10 miles distant bearing south. 1.30 a.m., light S. by E. wind, tacked to eastward. Current since noon of 19th has been setting about 1½ knots to N., N.W. and westward. 3 a.m., light air again from E.S.E., tacked to southward; freshened a little and weathered the Button by about a mile. About 5.30 a.m., all staysails and fore-topmast and topgallant stunsails set and Jamie Green. Yards sharp up; wind freeing a little as we neared Anjer. At 7 a.m., hoisted our number, etc., also sent our report, 21 days from Foochow, ashore and letter for owners (one dollar for postage to master attendant), bought fruit, fowls, etc. Faint baffling airs from north-eastward and calms. Current setting to S.S.W. about 2 knots. Noon, lat. 6° 8′ S., long. by(bearing) 106° 46′ E. P.M., light breeze from N.N.E. and clear. 6 p.m., light breeze from N.E. hauling to eastward. All possible sail set. 11 p.m., West Point of Princes Island bearing south about 12 miles, clear moonlight; still passing vessels bound northward, twenty at least in the Strait.

Thursday, 21st June, 1866.—Fine steady moderate breeze and clear from E.S.E. Employed as formerly and making a ringtail of No. 4 extra, 76 yards of canvas. Noon, lat. 7° 50′ S., long. 103° 11′ E.

Friday, 22nd June, 1866.—Some wind veering from E. to S.E. and back to E. Altered the course more southerly to pass south of Keeling Island and get more wind if possible. Employed as above and scraped pitch off deck along the seams, etc. Noon, lat. 10° 5′ S., long. 100° 16′ E. Distance 215 miles. P.M., holystoned the topgallant forecastle head and thoroughly washed the decks ready for varnishing tomorrow—weather permitting. Midnight, steady E.S.E. wind and clear.

Saturday, 23rd June, 1866.—Same weather. Varnished the deck. Carpenter finishing the gangway ladders, rope carving along the edge. Noon, lat. 12° 57′ S., long. 96° 48′ E. Distance 290 miles. P.M. till 3, very unsettled with rain. About 3.30 p.m., fresh S.E. trade wind again came away. Watch setting up topmast rigging.

Sunday, 24th June, 1866.—Strong squalls from S.S.E. and very confused S.W. sea. Ship pitching and surging to leeward considerably. In all small sails and stunsails 8 a.m., leechrope of main topgallant sail gave way and split the sail. In main royal and shifted main topgallant sail. 10.30 a.m., set them again and fore-topmast stunsail. Noon, had set fore and mizen royals and topgallant staysails. Noon, lat. 14° 9′ S., long. 91° 51′ E. Distance 280 miles. P.M. and midnight, fresh breeze and high southerly sea.

Monday, 25th June, 1866.—Fresh steady S.E. by S. trades and high southerly sea; less squally; set topgallant stunsails and main topmast and lower stunsails; sea less from ahead. Watch about the rigging as formerly. Carpenter putting on the iron band again round rudder and sternpost head; have to ease the wood at afterpart to let the hoop down and to work more freely. Sailmaker making a ringtail. Noon, lat. 14° 57′ S., long. 86° 30′ E. Distance 317 miles. P.M., same weather. Shipping water over all these two days past—chiefly at the end.

Tuesday, 26th June, 1866.—Same weather, less wind. All possible sail set. Employed as above. Carpenter making stunsail yards; having carried away two topmasts, one royal and one topgallant stunsail yard (two latter when in use for Jamie Green). Noon, lat. 16° 11′ S., long. 81° 3′ E. Distance 330 miles. P.M., wind more aft.

Wednesday, 27th June, 1866.—Wind light from E. and E. by N. Set starboard stunsails and ringtail and watersail under it and laced mizen staysail to outside of lower stunsail. Noon lat. 17° 23′ S., long. 76° 28′ E. Distance 270 miles. P.M., same weather.

Thursday, 28th June, 1866.—Fine clear weather and light trade wind from E. and E.S.E., hauling to S.E. All possible sail set to best advantage. Employed as formerly reserving lower and topmast rigging. Set up topmast and topgallant rigging and re-rattled the rigging—not particularly till round the Cape. Carpenter making grating for fore-scuttle doorway. Sailmaker giving the ringtail 4 feet more roach in the foot—6½ feet in all. A topgallant stunsail set instead of it meantime. Noon, lat. 18° 30′ S., long. 72° 40′ E. Distance 230 miles. P.M., again bent the spare flying-jib for a jib topsail set on fore-royal stay well up.

Friday, 29th June, 1866.—Same weather. Similarly employed. Noon, lat. 19° 51′ S., long. 68° 18′ E. Distance 255 miles. The other evening the steward in opening a drawer found a box of matches had ignited and burned some paper near it. The end of the box had been nibbled by a rat and no doubt caused the matches to ignite—the closeness of the drawer probably smothered the fire—but this shows how fire may originate.

Saturday, 30th June, 1866.—Same light trade wind from eastward and clear weather. Sea getting smoother. Employed about the rigging, greased masts, cleaned steering-screw, etc., cleaned brass work, oiled yard trusses and steering-gear, etc., as usual on Saturdays—weather permitting. Noon, lat. 21° 19′ S., long. 63° 51′ E. Distance 270 miles.

Sunday, 1st July, 1866.—Wind lighter from E. by N. and smooth sea. Noon, lat. 22° 42′ S., long. 60° 30′ E. Distance 205 miles. P.M., same weather, sky very clear.

Monday, 2nd July, 1866.—Same weather. Employed about the rigging serving over ends of splices of heavy lower braces rove after leaving Anjer. Rove Europe main braces, the Manila braces were too heavy to work well through the bulwark. Sailmaker making a fore-royal stunsail. Carpenter making shelf to go between after fife-rail stanchions 8 inches above the deck for mess skids being opposite the galley door. Noon, lat. 24° 28′ S., long. 57° 18′ E. Distance 205 miles.

Tuesday, 3rd July, 1866.—Light breeze from north-eastward and clear weather. Noon, lat. 25° 52′ S., long. 54° 20′ E. Distance 193 miles. P.M., less wind and more northerly.

Wednesday, 4th July, 1866.—Heavy dew. Very clear till about 4 a.m. very light N.N.E. wind and sky overcast. 5.30 a.m., sudden shift of wind to westward, moderate with rain till 10 then very light S.W. wind. Noon, moderate with rain. Lat. 26° 39′ S., long. 52° 1′ E. Distance 120 miles. P.M., freshening from S.S.W., all sail set. Employed as yesterday and drying stunsails, etc. Unbent main topmast and large lower stunsails and ringtail, foresail bonnet, watersail, etc. Unrigged boom and made all secure about decks. Midnight, continued clear and moderate from S.W.

Thursday, 5th July, 1866.—Wind hauling more westerly. Employed about the rigging and putting away aforesaid sails. Made a new fore-hatch tarpaulin to put on under two others. Repaired main and fore topgallant sails, both split recently by rope breaking just below the inside head cringle. Mainsail went the same way (rope seems rather light); bent said sails again and put away the best suit. Towards noon, light wind from S.W. by W., and clear weather. Long swell from westward. Noon, lat. 26° 57′ S., long. 48° 30′ E. Distance 190 miles.

Friday, 6th July, 1866.—Same weather till about 8 a.m. Wind very light and hauling to south and S.S.E. Set port stunsails. Employed as yesterday. Carpenter making a grating for deck at after part of quarter hatch. Sailmaker making the fore-royal stunsail. Unscrewed and cleaned bobstay and bowsprit shroud, setting up screws (fourth time this voyage), and oil them well with castor oil, also other screws, etc., requiring to keep all in good working order. Noon, lat. 27° 1′ S., long. 46° 36. E. Distance 110 miles P.M. and midnight, very light easterly wind.

Saturday, 7th July, 1866.—Same weather and wind, hauling to E. and E. by N. Noon, lat. 27° 58′ S., long. 44° 25′ E. Distance 132 miles. P.M. and midnight, same weather and cloudy.

Sunday, 8th July, 1866.—Same weather. Noon, lat. 29° 23′ S., long. 41° 21′ E. Distance 180 miles. P.M., better breeze and hauling to N.E.

Monday, 9th July, 1866.—2 a.m., wind hauling to N.N.E., sky over-cast, lightning in the westward. In after stunsails. 3.30 a.m., rain. In all stunsails and small sails. 4 a.m., fresh squall from N.W. In main royal and mizen topgallant sail and flying-jib. 4.30 a.m., light wind. 5 a.m., made sail again, heading south-westward, showery and light baffling W.N.W. wind, 10 a.m., wind shifted to S.W.; tacked to W. by N.; clearing up. Noon, lat. 30° 52′ S., long. 38° 58′ E. Distance 157 miles. P.M., drying stunsails and other work. Midnight, moderate southerly wind; all sail set.

Tuesday, 10th July, 1866.—A little squally, in royal staysails when needful.

Carpenter fishing a fore-topmast stunsail boom; he let in a piece of teak, it will be rounded with fine twelve thread stuff, having no hoops, and he is fitting a rope strop to answer for yardarm iron—the same as fitted for ringtail boom. Watch seizing off the lower lanyards snug, etc., and making sennet. Noon, lat. 32° 24′ S., long. 34° 58′ E. Distance 222 miles. P.M., clear and fine breeze from S. by W. 4 p.m., passed ship *City of Bombay* from Calcutta to London 47 days out. She was 6 miles ahead at noon. Midnight, less wind from S.S.E.

Wednesday, 11th July, 1866.—Continued fine weather and light wind, hauling to S.E. and E. Noon, lat. 33° 34′ S., long. 30° 54′ E. Distance 227 miles. P.M., wind hauling more to N.E. and N.N.E.; in after-stunsails at 10 p.m. Midnight, fresh breeze and clear.

Thursday, 12th July, 1866.—Wind freshening rapidly from 2 to 5 a.m. In royals and flying-jib. 5.30 a.m., much less wind, hauling to N. and N.W. Set all sail on a wind. Noon, light W.N.W. wind, smooth water and clear weather. Noon, lat. 34° 46′ S., long. 27° 16′ E. Distance 190 miles. 2 p.m., tacked to N.W.; light W. by S. wind. Watch employed making sennet, tapering ends of lower lanyard mats ready to go on by and by. Stoppered the leach-rope of fore-topsail which was gone at starboard clew (halliards should be eased just before rain or heavy dew if they have been sweated tight up in dry weather). Carpenter began to raise after-grating and binnacle 4 inches off the deck; preparing the teak and carving, etc. Midnight, very light westerly wind. Tacked to southward having got soundings in 75 fathoms, which agrees with reckoning.

Friday, 13th July, 1866.—Towards 9 a.m., wind freshening a little and hauling to northward; set fore-topmast and topgallant stunsails (jib topsail on fore-royal stay and Jamie Green set as usual). Noon, lat. 35° 17′ S., long. 24° 50′ E. Distance 131 miles. Experienced about 36 miles westerly current in the 24 hours. 4 p.m., wind N.W.; stunsails in. A barque to leeward and ship to windward, passing them fast, other vessels seen from aloft. Midnight fresh breeze from W.N.W. and confused sea.

Saturday, 14th July, 1866.—1 a.m., ship almost unmanageable in the strong current. She came round against the helm, and gathered dangerous sternway; braced round and trimmed on port tack. Till 5 a.m., strong gusts and confused sea; topgallant sails in. 5.30 a.m., it suddenly moderated; set all sail on wind. Noon, lat. 36° 1′ S., long. 21° 46′ E. Distance 159 miles. Experienced 52 miles of W.S.W. current in the 24 hours. P.M., wind very light then calm; clear weather. Shifted fore and main upper topsails; rope gone at clews. 6 p.m., faint air from N.E.; set stunsails. Midnight, light breeze, clear, and heavy dew.

Sunday, 15th July, 1866.—Clear weather and light northerly wind, hauling to N.W. In stunsails, etc., as the wind hauled. Signalised British ship *Tantallon Castle* from Calcutta to London, 45 days out; we came up to and passed her easily. Noon, lat. 35° 53′ S., long. 20° 12′ E. Distance 82 miles. Current 8 miles east. P.M., faint N.W. and W.N.W. winds. 4 p.m., tacked to northward. 6 to 7 p.m., nearly calm then faint air from S.E., ship barely steering, freshened towards 9 p.m.,; set stunsails. Midnight, barometer falling; getting cloudy and misty, dew falling, confused sea. In royal stunsails.

Monday, 16th July, 1866.—1 a.m., in topgallant stunsails, skysail, wind hauling to E.N.E. and N.E. 3.30 a.m., sky clearing and steady moderate wind; set topgallant stunsails and skysail again. 7 a.m., wind hauling more northerly. In after and lower stunsails. Passing several vessels going same way. Noon, lat. 35° 40′ S., long. 17° 30′ E. Distance 130 miles. 14 miles of current E. by S. P.M., wind hauling more and freshening. In stunsails, staysails, Jamie Green, skysail and fore and mizen royals. 4 p.m., fresh breeze from N.N.W.; getting cloudy. Barometer which has been from 30·20 to 30·60 since passing Mauritius now 30·00. 8 p.m., barometer 29·98. More wind and sea, topgallant sails and flying jib in.

Tuesday, 17th July, 1866.—Wind increasing. Crossjack and spanker furled to ease her in pitching. Mizen staysail set. At daylight, less wind. Set topgallant sails and spanker. 10.30 a.m., set main royal and flying jib. 11.45 a.m., dark cloud rising ahead. Up mainsail, in crossjack and mizen staysail, main royal and flying jib. Noon, wind headed off to S.W. and S. Up foresail and wore round with light westerly wind and cloudy weather. On port tack heading N. Noon, lat. 36° 3′ S., long. 14° 23′ E. (by account). Distance 160 miles. P.M., wind increased to a moderate breeze and head sea falling. Set all sail gradually and fore-topmast stunsail.

Wednesday, 18th July, 1866.—Light westerly and S.W. winds. Watch preparing the mats; repairing upper topsails, etc. Noon, lat. 34° 35′ S., long. 12° 21′ E. Distance 133 miles.

Thursday, 19th July, 1866.—Same wind and generally clear weather. Passing four vessels fast. 10 a.m., wind hauling to south; set starboard stunsails. Employed getting all coal up out of coal hole; put two tons in the shifting box and about 1 ton in galley locker, sheep pen and pig house (both pigs having been killed, one yesterday and the other ten days previously). Carpenter making a grating cover for sail cabin hatch. I will condemn the present heavy teak cover. Also making a new fancy canvas cover of No. 2 to go under the present canvas cover, to be thoroughly watertight when secured down. Noon, lat. 32° 54′ S., long. 10° 55′ E. Distance 133 miles. 2.30 p.m., wind hauled to S.S.E., gybed spanker and shifted fore and lower stunsails over; bent and set fore-royal stunsails. Drizzling showers with passing light squalls; wind veering from S. by E.; all port stunsails in.

Friday, 20th July, 1866.—2 a.m., in a slight shower which rose on starboard quarter, wind for 2 or 3 minutes came from N.E.; kept off W. by N., but it backed immediately to S.E.; took in royal stunsails and jib topsail, skysail and royals and downed staysails. 2.15 a.m., set royals, skysail and staysails, and trimmed yards; wind E.S.E. Daylight, set all possible sail, wind still veering from E. by S. to S.E. with passing clouds. Employed as yesterday. Carpenter finished hatch cover. Noon, lat. 31° 10′ S., long. 7° 31′ E. Distance 188 miles. P.M., took in, set and trimmed sail as necessary.

Saturday, 21st July, 1866.—Increasing head sea. Wind gradually headed ship off to N.N.W and N.W. Noon, light baffling N.N.W. wind veering with every cloud. Lat. 29° 20′ S., long. 2° 45′ E. Distance 271 miles. P.M., calms and baffling airs, tacking, etc., as necessary.

Sunday, 22nd July, 1866.—Wind all round the compass, clouds generally moving from northward, sky partially clear. Noon, lat. 29° 27′ S., long. 2° 4′ E. Distance 40 miles. P.M., calms and baffling.

Monday, 23rd July, 1866.—Same weather. Tacking and trimming yards, etc. Sent up one of the Foochow spare spars for main-topmast stunsail-boom—the fished boom having given way. Bent the second fore and main upper topsails; put away the best. Noon, lat. 28° 56′ S., long. 0° 57′ E. Distance 75 miles. P.M., calms and baffling airs; very clear and smooth sea.

Tuesday, 24th July, 1866.—Faint airs from westward. Steering to make northing to reach limits of S.E. trades as soon as possible. Watch putting on lower lanyard mats; swiftering and sparing the rigging ready for rattling down. Carpenter making eight teak capstans bars for a fixed rack on after part of house, also made a proper water funnel—lead nozzle; copper wanted. Main skysail very thin making another of four cloths No. 3 in middle, wings of No. 5, taking the bonnet of foresail as we have no other light canvas. Cleaned the coal hole thoroughly and gave a second coat of pine oil over iron also. Noon, lat. 27° 31′ S., long. 0° 27′ E. Distance 85 miles. P.M., wind hauling to S.W.

Wednesday, 25th July, 1866.—Wind hauling to south and S.E. freshening a very little; all possible sail set, ringtail, water-sails, save-alls, mizen-staysail wing to lower stunsail, etc. Found the ship is again on even keel. Using the water must lighten her aft, as, though we put the Manila and Europe tow ropes in the lower forecastle the other day from my cabin, we brought fully 2 tons of coal from coal hole to forepart of quarter-hatch in the shifting box, therefore moved said box aft alongside skylight starboard side to trim a little by the stern. Noon, lat. 26° 31′ S., long. 0° 12′ W. Distance 76 miles. P.M., wind freshening to moderate breeze and cloudy, veering with every cloud from S.S.E. to E. Yards trimmed for S.E. wind and edging off north as required.

Thursday, 26th July, 1866.—Wind falling light at times. Employed rattling lower and topmast rigging. Hitching the separate strands round the shrouds and laying up again between the shrouds for snugness. Carpenter at fancy capstan bars, sailmaker at the new skysail, and one hand making the fringe of sail cabin cover. Noon, lat. 24° 22′ S., long. 3° 20′ W. Distance 220 miles. Water when pumping ship in dog watch comes up very rusty with the kentledge and one scupper abreast is too small to discharge what pumps throw; a larger scupper and hose would keep decks clean. P.M., light baffling winds and cloudy.

Friday, 27th July, 1866.—Wind veering all round the compass, and calm; at times airs prevailing from eastward. Noon, lat. 22° 18′ S., long. 5° 32′ W. Distance 180 miles.

Saturday, 28th July, 1866.—Same weather. Strange winds where the S.E. trades usually prevail. Bent the new skysail—the other of No. 5 is as thin as paper in all respects. Noon, lat. 20° 13′ S., long. 7° 46′ W. Distance 180 miles.

Sunday, 29th July, 1866.—Same weather and showery. Noon, lat. 18° 58′ S., long. 9° 2′ W. Distance 105 miles. 6 p.m., began to freshen from east. Midnight, steady, moderate E.S.E. trade wind.

Monday, 30th July, 1866.—Wind hauled to S.E., sky clearing with steady trade wind. 8.30 a.m., a three-masted schooner *Mondeco*, bound south, signalled, "Can you spare provisions?" We replied, "No." She proceeded close hauled to southward. Employed rattling down, tarr'ng, etc. Carpenter putting down carved teak base for binnacle and athwartships to raise after gratings to same level, 4 inches above deck. Noon, lat. 16° 30′ S., long. 10° 18′ W. Distance 165 miles.

Tuesday, 31st July, 1866.—Same wind and weather. Noon, lat. 13° 5′ S., long. 12° 36′ W. Distance 240 miles.

Wednesday, 1st August, 1866.—Same weather. Employed black painting ties, topsail sheets, etc., and seizing on short spars 2 inches higher than every ratlin—to remain till near arrival to keep all in best order. Poured about 1 gallon of pine oil into rents of spare spars, also jibboom that is out, to keep wood fresh and sound. Will turn the spars over in about a week and fill rents on other side. Noon, lat. 8° 46′ S., long. 13° 58′ W., Distance 270 miles. P.M., sighted Ascension Island. 4.30 p.m., abreast S.E. points, distant 3 miles. Saw three ships running during the day, left them all fast. Towards midnight, wind getting light.

Thursday, 2nd August, 1866.—Same weather. Employed scrubbing inside of boats; pumping all fresh water out of port tank (it just filled other tank and all the small casks), then filled the port tank with salt water to trim ship by the stern, as we can't have the shifting box full of coal close aft while cleaning ship. Also finishing odds and ends about the rigging and putting on mats and battens after tarring. Towards noon, wind very light and sultry weather. Noon, lat. 5° 49′ S., long. 15° 40′ N. Distance 208 miles. Steering for 21° 22′ W. at crossing the line. Wind very light from S.S.E. to S.E. by S.

Friday, 3rd August, 1866.—Light steady S.S.E. wind and clear. Employed painting inside of boats. Had all the bottom boards unshipped—painted them, under and upper sides, ditto side, puddings and belts thoroughly. Noon, lat. 3° 35′ S., long. 17° 57′ W. Distance 191 miles.

Saturday, 4th August, 1866.—Same wind and weather. Carpenter finished the carved front for after gratings—did not fix them in place till ship is clean. Got up eight empty small tanks out of store-room and painted them and stowed the tea (12 chests and 22 half chests), which has been in the cabin locker-heads to trim the ship by the stern. Stowed some eight cans of ginger and twelve boxes of tea in starboard water-closet, etc. Holystoned under the long boat, to oil the deck while the boat is standing keel down. Scraped the lignum vitae fair-leads and upper deadeyes and oiled them. Noon, lat, 1° 28′ S., long. 19° 54′ W. Distance 170 miles. P.M., a barque in company dropping astern fast.

Sunday, 5th August, 1866.—Same weather. Steering right before the wind. Noon, lat. 0° 20′ N., long. 22° 1′ W. Distance 170 miles. 67 days to the line from the Foochow River.

Monday. 6th August, 1866.—Same light steady S.S.E. winds and clear. Employed turning boats down on the skids again;; stowed the small tanks on a plank on the skids within the after boats, stowed the pighouse in its place under the large boat forward, after pine oiling the deck and anchors under said boat. Scraped all the spots of pitch, tar grease, etc., off the paint-work aloft and about the bulwarks, rails and waterways, etc., then commenced scrubbing the bright work and paint work with sand. Broached the full tank of fresh water (2050 gallons) today. Noon, lat. 2° 59′ N., long. 24° 31′ W. Distance 220 miles. 1 knot current in favour.

Tuesday, 7th August, 1866.—Same weather. Carpenter repairing bulwarks where stove in, port side of fore rigging and preparing bottom boards of teak for the sheep pen. Noon, lat. 6° 18′ N., long. 26° 12′ W. Distance 221 miles. 1 knot of current to northward. P.M., scraped rust blisters to the bare iron of mizen and mainmast and rubbed clean and smooth with sand-stone, also a few spots on foremast and bowsprit, then coated with white zinc which when dry will be smoothed along edges of blister marks with pumice-stone before painting mast colour. Midnight, nearly calm; sky overcast.

Wednesday, 8th August, 1866.—Light wind hauling to westward. 5 a.m., braced nearly sharp on port tack; stunsails set forward. Employed scrubbing bright work and paint work with sand and pumice-stone; inside bulwarks, etc. Carpenter put new teak bottom on the sheep pen. Noon, lat. 8° 33′ N., long. 26° 58′ W. Distance 145 miles.

Thursday, 9th August, 1866.—Light and moderate winds from westward. Sky generally clear. Similarly employed, and carpenter taking off the sheathing of starboard fore and port mizen channels. Noon, lat. 9° 59′ N., long. 27° 7′ W. Distance 85 miles. Midnight, wind moderate from W.N.W.

Friday, 10th August, 1866.—Wind baffling, still scrubbing with sand. Scraping smooth and pumice-stoning. Carpenter doing carving on handspike rack. At 4 a.m., wind died away to a calm. 5 a.m., light misty air from N.W. hauling to north. 6 a.m., tacked with light showery weather and N.N.E. wind which freshened, promising to be the N.E. trades. Noon, lat. 12° 42′ N., long. 28° 28′ W. Distance 185 miles. Passed two ships running south. P.M., wind rather unsteady.

Saturday, 11th August, 1866.—Wind veering from N. by E. to N.E., from fresh breeze to a calm. Finished scrubbing inside and scrubbed the bulwarks outside to a plank below the covering board. Carpenter taking off all channel sheathing on starboard side. Will leave port fore and main till across N.E. trades to help her when side is down. Noon, lat. 14° 2′ N., long. 31° 7′ W. Distance 184 miles. P.M., washed down thoroughly inside and out. Today commenced burning wood instead of coal to keep paint work clean.

Sunday, 12th August, 1866.—Same weather. Noon, lat. 15° 56′ N., long. 33°

12′ W. Distance 162 miles. P.M., a barque in company dropping astern. Slight showers occasionally. Same veering light, north-easterly winds.

Monday, 13th August, 1866.—Wind steadier and clear weather. Commenced to oil the bright work and paint inside ship, first the white panels. Noon, lat. 17° 30′ N., long. 34° 59′ W. Distance 154 miles. Carpenter on small jobs in order not to make chips—lower ends of carved ornaments on house too fragile, shortened them a little.

Tuesday, 14th August, 1866.—2 to 6 a.m., showery; paint mostly set fortunately. Till 9 a.m., scrubbing and scraping gratings, ladders, bucket racks, etc. Sky clearing with moderate E. by N. wind. Washed down and went on painting; gave bright work a second coat of raw oil rubbed on thin; and panels, etc., a second coat of white. (Mean to give everything two coats of paint and bright work three coats of oil, and after decks, etc., are cleaned, copal varnish the bright work.) Noon, lat. 20° 13′ N., long. 36° 12′ W. Distance 163 miles. P.M., gave all three boats second coat of white outside, also doing second coat of white and first of green inside the bulwarks.

Wednesday, 15th August, 1866.—Light steady N.E. trade wind. Oiling and painting. Carpenter made cement of fine dry lime and tar and refilled over bolt-heads of covering board. Noon, lat. 23° 33′ N., long. 37° 18′ W. Distance 210 miles.

Thursday, 16th August, 1866.—Very light easterly wind and clear, warm weather. Carpenter took off remainder of channel sheathing—port side. Watch painting and oiling. Painted waterway plank blue mixed with zinc. Gave the masts and mastheads etc., first coat nearly flesh colour with patent and white zinc mixed. Noon, lat. 26° 49′ N., long. 38° 11′ W. Distance 202 miles.

Friday, 17th August, 1866.—Finished painting and oiling inside, and most of masts; sails flapping too much to finish all, being nearly calm. Scraped the channels and washed outside below covering board. Noon, lat. 28° 7′ N., long. 38° 39′ W. Distance 82 miles. P.M., painted black ropes, inner part of guys, lower-brace pennants lead colour. 9 p.m. to midnight, calms and baffling airs. Hauled yards round as necessary.

Saturday, 18th August, 1866.—Baffling and calm till 6.30 a.m. then clear and freshening air from S.W. and W.S.W. Set stunsails. Finished painting the masts aloft before 8 a.m. Wind hauling to west and N.W., steering N.E. ½ N., corrected compass. In stunsails, etc., when necessary. Employed painting outside from rail to two planks below channels all round. Noon, lat. 28° 58′ N., long. 38° 40′ W. Distance 55 miles. 4 p.m., finished painting; commenced to break sand for cleaning decks. Put the newly painted small casks (for vinegar, molasses, etc,) below in lower forecastle. Put hold ladder and stage planks across the stern again, stowed there some ten days ago, with broken booms, etc., partly resting on middle of taffrail, ends on two top-gallant stunsail booms lying on the bumpkins and outer edge of quarter chops of taffrail. Broke up the shifting coal box and otherwise prepared for cleaning decks, put the coal forward into coal locker. Midnight, faint north-westerly airs and clear weather, sea smooth.

Sunday, 19th August, 1866.—Calms and faint airs. Noon, lat. 30° 36′ N., long. 37° 43′ W. Distance 108 miles. P.M. and midnight same weather.

Monday, 20th August, 1866.—Same weather. Tacked, squared away, etc., as necessary. Commenced holystoning decks. Carpenter replacing two large pieces of carved work washed away when caught aback one night in south latitude. Cleaned and painted inside of house. Noon, lat. 30° 56′ N., long. 37° 26′ W. Distance 32 miles.

Tuesday, 21st August, 1866.—Same weather. 2 a.m., tacked to N.N.W., light breeze from N.E. clear weather, a little swell from N.W. Noon, lat. 31° 29′ N., long. 37° 21′ W. Distance 36 miles.

Wednesday, 22nd August, 1866.—Same weather. Watch going over decks with large and small holystones the second time. Carpenter fitting stands of sidelights abaft mizen rigging with a rake out to show light right ahead of courses and all ropes. Noon, lat. 32° 40′ N., long. 39° 3′ W. Distance 114 miles.

Thursday, 23rd August, 1866.—Light baffling winds and calms. Much time employed in trimming sails, otherwise holystoning decks, scrubbing gratings, bright bucket rack and buckets (12 in number) gun carriages, etc. Carpenter preparing teak to put a new edge bottom round manger at fore part of main hatch, the present one is too light and shattered. Noon, lat. 33° 9′ N., long. 39° 38′ W. Distance 44 miles.

Friday, 24th August, 1866.—Same weather and showers. Washing down rails, bulwarks and decks, etc., twice over, and rope's ends overboard. Noon, lat. 34° 56′ N., long. 39° 17′ W. Distance 108 miles. 6 p.m., tacked again to north, wind N.E. by E. freshened to a smart breeze, all staysails in and jib topsail. Towards midnight light breeze, set jib topsail, mainroyal, and topgallant and middle staysails as usual when on a wind. Had the decks, etc., all clean for oiling.

Saturday, 25th August, 1866.—Very light N.E. wind and fine clear weather. A.M., washed decks and swept and wiped all thoroughly dry and lifted everything possible off the deck. 9 a.m., commenced with China wood oil, gave two good coats and a third where it would take it, then went over all again a fourth time with dry rags, rubbing fore and aft. Used 16 gallons of oil and one gallon of raw oil to finish. Finished all the decks at 4 p.m., had a splendid day of it. Carpenter at sundry small jobs aloft. Painted rudder-head, tiller and part of steering gear green, and all under the monkey poop brown, teak colour; also painted one waterway plank second coat of blue, and commenced to turpentine varnish the bright work, viz., after part topgallant forecastle, fore-scuttle stanchions, top of covering board, fife rails fore and aft, round masts, etc. Noon, lat. 36° 20′ N., long. 40° 1′ W. Distance 95 miles.

Sunday, 26th August, 1866.—Wind hauling to E.N.E. and east, very light. Noon, lat. 37° 34′ N., long. 40° 24′ W. Distance 76 miles.

Monday, 27th August, 1866.—Same weather. Set starboard stunsails forward and main topgallant and royal stunsails, all other sail set and Jamie Green and jib topsail. Employed scraping remainder of gratings, buckets, etc. Painted the spare large spars. Noon, lat. 38° 52′ N., long. 39° 56′ W. Distance 84 miles. 9 p.m., breeze began to freshen a little, fine clear weather and smooth sea. Midnight, wind S.S.W., going about 6 knots.

Tuesday, 28th August, 1866.—Same weather. Similarly employed, hope to finish bright work tomorrow. Spars aloft, stunsail booms and yards still to clean and oil. Steward this week past had to bake bread for all hands. We have three weeks' flour remaining at full allowance. He has cleaned the cabin after removal of tea, also pantry, etc. Noon, lat. 40° 16′ N., long. 38° 7′ W. Distance 124 miles. P.M., 7-knot breeze from W.S.W. 10.30 p.m., same from W. by N.

Wednesday, 29th August, 1866.—Light breeze and clear from west hauling to N.W. and N.N.E. Scraping, planing and oiling spars. Sent down main topgallant and royal stunsail booms, planed and oiled them and sent them up again. Same with other booms. Painted roof and inside of skylight and all my cabin. Noon, lat. 41° 54′ N., long. 34° 32′ W. Distance 193 miles. 6 p.m., wind shifted from W.N.W., to N by E. suddenly; in stunsails and braced up. Midnight, moderate north wind and cloudy.

Thursday, 30th August, 1866.—Wind hauling again to N.W., west and S.W. Similarly employed. Noon, lat. 43° 33′ N., long. 30° 46′ W. Distance 194 miles.

Friday, 31st August, 1866.—Same weather with rain. Noon, lat. 45° 2′ N., long. 27° 38′ W. (by account). Distance 162 miles. 3 p.m., still raining. Wind hauling to south and S.E. in gusts, in small sails. 7 p.m., wind from S.E. by E., getting light and clearing, made sail. 10 p.m., very light wind hauling again to south and S.S.W. 10 p.m., lat. by Pole star 45° 35′ N. (45° 48′ N. by account). Midnight, light and freshening S.W. wind. Moon and stars showing at times.

Saturday, 1st September, 1866.—Moderate wind, misty, showers from W.S.W. sky clearing towards 10 a.m. Noon, lat. 46° 39′ N., long. 24° 3′ W. Distance 176 miles. 2 p.m., wind hauled to west and W.N.W. Gibed and set stunsails port side. Sun showing, fine moderate breeze. Passed many outward-bound vessels. 7 p.m., cloud rose from north and wind suddenly veered from W.N.W. to N.E. by E., strong breeze and small rain. In stunsails and small sails and braced up. Midnight moderate and clearing from N.E. by N.

Sunday, 2nd September, 1866.—Moderate and light N.E. north and N.W. wind still hauling to west and S.W. Stunsails, etc., set accordingly. Noon, lat. 47° 22′ N., long. 19° 42′ W. Distance 185 miles. P.M., fresh gusts from south and S. by E., and rain. In stunsails as the wind freshened.

Monday, 3rd September, 1866.—1 a.m., wind still freshening with rain and squalls. In royals, fore topmast stunsail and flying jib at 2 a.m. 5 a.m., in fore and mizen topgallant sails. 5.30 a.m., in main topgallant sail. 6 a.m., in crossjack and spanker and eased upper topsail halliards. 6.30 a.m., rope fore tack hauling part dragged the capstan out of the broken sockets; secured again with chain fore tacks to bitts. Gusts very strong. 7 a.m., called all hands, reefed the mainsail and set it again. Wind began to back again to S.S.W. and S.W. 9 a.m., more moderate from west, set all plain sail, kept reef in mainsail. Noon, lat. 48° 58′ N., long. 15° 42′ W. Distance 192 miles. P.M., set all starboard stunsails forward and starboard main topgallant stunsail. Watch got anchors on the bows and chains bent all clear. Barometer at 9 a.m., to 5 p.m., 29·79° steady. Stopped falling but has not risen at all.

Tuesday, 4th September, 1866.—Moderate S.W. wind and rain, thinning off at times. All sail set. 7 to 10 a.m., very light wind; again freshened with showers. Sun showing at times. Noon, lat. 49° 26′ N., (account 49° 22′), long 10° 20′ W. (account 10° 42′). Distance 212 miles. Towards sundown more wind and rain, gusts from S.W. and thick gloomy weather.

Wednesday, 5th September, 1866.—Same weather continuing. Wind veering from S.W. by S. to W.S.W. and back, with lightning in the west. Approaching Scilly Isles. In small staysails, skysail and topgallant stunsails. 1.30 a.m., saw Bishop and St. Agnes Lights. 2.50 a.m., St. Agnes, distant about 10 miles, bore north by compass, ship's head E. by S. ½ S. 5.30 a.m., sky cleared. Set all possible sail. Barometer 29·45°, began to rise. 8.25 a.m., Lizard Lights about W.N.W. 11 miles. Noon, nearing Start Point. 30 min. p.m., the lighthouse bore north, distant 3 miles. Hoisted our number. A ship since daylight has been in company on starboard quarter. *Taeping* probably. 4.15 p.m., Portland Lights north, distant about 6 miles. Have been going 14 knots; royal stunsails and all flying kites set, wind strong from W.S.W., hazy but clear overhead. Before 6 p.m., got anchors over, Jamie Green and jib topsail unbent, and otherwise all clear forward. 6 p.m., barometer 29·59, rising very slowly. 7·25 p.m., St. Catherine's Light north one mile. In all small sails except fore topmast stunsail. 9.45 p.m., Owers Light north 4 miles. 12.30 a.m., Beachy Head Light north 5 miles.

Thursday, 6th September, 1866.—30 min. a.m., gradually reduced sail as we neared Dungeness. 3 a.m., Dungeness Light N.E. 8 miles. Upmain sail, in main royal fore and mizen topgallant sails, flying jib and up foresail. Sent up rockets and burned several blue lights. 4 a.m., in main topgallant sail and hove to abreast of Dungeness Light, distant 1½ miles. Continued to make signals for a pilot. 5 a.m., saw the *Taeping* running and also signalling. Bore up lest they should run eastward of us and get pilot first. Seeing us keep away, they hove to and we again hove to. 5.30 a.m., saw two cutters coming out of Dungeness Roads. 5.40 a.m., kept away so as to get between *Taeping* and the cutters. 5.55 a.m., rounded to close to the pilot cutter and got first pilot, and were saluted as first ship from China this season. I replied, "Yes, and what is that to the westward? We have not room to boast yet." Thank God we are first up channel and hove to for a pilot an hour before him. 6 a.m., kept away for South Foreland; set all plain sail; were immediately followed by *Taeping*. They set also topmast, topgallant and lower stunsails one side; wind slightly quartering. We kept ahead without the stunsails or would also have set them. *Taeping* neared us a mile or two but was a mile astern when he had to take in his stunsails (he had shifted them to the port side when hauling up through the Downs). Hoisted our number abreast of Deal. We were then fully one mile ahead of *Taeping* and kept so till obliged to take in all sails and take steamer ahead. *Taeping's* tug then proved much better than ours and soon towed past us. I thought of taking another boat but found there would be no need as far as docking was concerned, as we could reach Gravesend two or three hours before it would be possible to go on account of tide, therefore I saved the £10 or £12 asked by boats. *Taeping* reached Gravesend 55 minutes before us. We avoided anchoring by getting a tug alongside to keep us astern. Proceeded with first tug ahead as the flowing tide gave us sufficient water to float, thus reached Blackwall and East India dock entrance at 9 p.m. They could not open the gates till tide rose higher. 10.23 p.m., hove the ship inside the dock gates. *Taeping* had preceeded us up the river, but having further to go, did not reach the entrance of London docks till 10 p.m., and drawing less water than us; also the dock having two gates, they got her inside outer gate, shut it and allowed the lock to fill from the dock, then opened the inner gate, so she docked some 20 minutes before us—the papers have it half-an-hour for the sake of precision—a strong westerly gale since 8 p.m.

APPENDIX I.

Spar Plan of "Cutty Sark."

BOWSPRIT.	Jibboom (extreme length)	60 feet
FOREMAST.	Extreme length (deck to truck)	129·9 „
	Lower mast (deck to cap)	61·9 „
	Lower mast (masthead)	14 „
	Topmast	48 „
	Topgallant mast	26 „
	Royal mast	17½ „
	Fore yard	78 „
	Lower topsail yard	68 „
	Upper topsail yard	64 „
	Topgallant yard	48 „
	Royal yard	38 „
MAINMAST.	Extreme length (deck to truck)	145·9 feet
	Lower mast (deck to cap)	64·9 „
	Lower mast (masthead)	14 „
	Topmast	48 „
	Topmast (masthead)	9 „
	Topgallant mast	26 „
	Royal mast	15 „
	Skysail mast	14½ „
	Main yard	78 „
	Lower topsail yard	68 „
	Upper topsail yard	64 „
	Topgallant yard	48 „
	Royal yard	38 „
	Skysail yard	34 „
MIZENMAST	Extreme length (deck to truck)	108·9 feet
	Lower mast (deck to cap)	55·9 „
	Lower mast (masthead)	11 „
	Topmast	38 „
	Topgallant mast	19½ „
	Royal mast	13½ „
	Crossjack yard	60 „
	Lower topsail yard	54 „
	Upper topsail yard	48 „
	Topgallant yard	39 „
	Royal yard	33 „
	Spanker gaff	34 „
	Spanker boom	52 „
	Topmast stunsail boom	47 „
	Outer end of flying jibboom to end of spanker boom	280 „

APPENDIX J.

Letter from Captain Joseph Wilson, last owner of "Challenge."

SUNDERLAND, *May*, 17, 1892.

DEAR SIR,—I bought the *Golden City* in 1866, in London, during a sort of monetary panic. Happening to have some business on board a Calcutta ship in the East India Dock, I passed over the deck of this ship, and observed what a magnificent main deck she had. I was informed by the ship-keeper that she had to be sold by auction. This man had been employed on my own ships and knew me, showed me all over her. I saw that she was opened out for the emigration inspectors and had just come out of dock, passed and ready for emigrants. The time I had being short, I merely took the broker's name and left. Knowing the owners and all the circumstances, accounted for the ship being on the market. All the information I got was, "confounded pick-pocket," "weak," "won't sail." One man said he "wouldn't be paid to own her." However, I walked down at six o'clock the next morning with a couple of wax candles in my pocket and a sort of instrument for constitutional purposes; found all sound and good and at 10 o'clock I bought the vessel, intending her for 2nd class work. On arrival at Shields, cut listings out of her to the water line. The frame I found to be as sound as when built and firmly cross-braced by iron diagonal strips. A lot of blocks and other things were found on over-hauling her with the name *Challenge* on them. I looked up Lloyd's and found that that ship had been classed in the book, and after some little bother got all unravelled and proved that this was the same old American clipper *Challenge* and consequently entitled to classification on survey. On examination, all was right but the fastenings of the outside plank, which was only spiked on. A lot of work had been done and money spent previously. I re-renailed her throughout and extra bolted her; shortened upper spars and sent her to sea.

She did very well first voyage; was difficult to keep in trim, but made a good run home. I recollect seeing her log from New Orleans; there were several days with over 300 miles. I then sent her to India, and she was principally employed in the Bombay and Java trades. She ran out to Bombay in 71 days; thence to Rangoon was 11 days. She made very good passages, but in general was not, with me, the very fast ship she had been originally. On a wind, going clean full, or all sails drawing she was hard to beat; she was a dangerous ship to stay, that is, in a strong wind and sea, and that was eventually the cause of her loss. Though to look at, in every way of a very handsome appearance both in hull and rig, no handsomer to be seen, yet in detail there were serious errors. The principal error was a hollow water-line; this was such a mistake; a good-sized pilot coble, in trying to get alongside in perfectly smooth water with ship towing only 7 knots, was upset by the curl of the wave caused by this hollow; a sea of 3 feet was always curling up between the stem and the fore rigging when going fast, and in ballast especially this effectually stopped her from being the very fastest vessel afloat. She was very stiff and rolled with a heavy cargo, but on the whole did her work well. On her last voyage to Java, in a heavy gale off the Cape of Good Hope, when running before it, and going 12 or 13 knots, a sea broke clean over her quarter, swept 7 men off the main deck, killed the captain on top of the house, took clean away the wheel and officer's house, one man at the wheel was saved by holding on to the mizen rigging 10 or 15 feet above the poop, all the officers in the house were gone but the third mate, my nephew; he was caught by the smashed beams and found very much injured, but recovered. The ship made no water and arrived out, but another captain had to be sent out. The ship lost her season and had to go on to Calcutta, and eventually made a most disastrous voyage. She was lost on the next—in consequence of her bad propensity in staying. She came up well enough, head to the wind, off Scilly got stern way; she came round but twisted off her rudder and reached over to the French coast, where she got among the islands off Ushant. The crew left her and a French gun-boat got hold of her, and a lot of wooden-shoed and wooden-headed fishermen, and instead of towing the ship stern foremost, took her by the bow, when she took a shear and went on a reef of rocks. Being so sharp on the bottom she upset and broke up off Aberbrache. This was the end of her, winding up with the loss of over £10,000 to me.

INDEX

INDEX

INDEX

INDEX

INDEX

INDEX